The triumph of Mathematics

VÍCTOR LAURIA

This book is dedicated to my wife Paula, my parents Eduard and Marisé, my brother Albert and my dear nephews Alejandra, Juan, Jorge, Paula, Pablo, Álex and Víctor

I want to thank the work of my friend Eduardo Lorente, who patiently reviewed all the chapters and helped me to improve many of them, the English translation that my friend Marta Conde disinterestedly corrected and the advice of my cousin Eduard Martorell for the correct edition of the book.

Title: The triumph of Mathematics
Author: Víctor Lauria
Date: March 2017
Cover design: Carles Marsal

Introduction

This book is a compilation of historical problems of Mathematics, arranged chronologically, that date between the sixth century BC (problem of the Pythagorean Theorem) until almost the end of the seventeenth century (problem of the impossibility of the sum of cubes). All these problems were solved a long time ago, but I took special care to present a simple solution for each of them (even if it was longer), trying that all those who love Mathematics can understand it without very sophisticated tools. My intention is to publish another book in the future with other 30 interesting problems that will cover the prolific eighteenth and nineteenth centuries.

Many proofs are not completely rigorous since that was not my intention; the goal of this book is to enjoy Mathematics. However, I am sure that the explanations are sufficient for any reader with the necessary mathematical foundation to find the reasoning completely valid. Each problem starts with a historical introduction that according to my opinion enriches the reading, and it ends with some observations that I hope will be of the reader's liking. As for its solution, I have tried to add figures and examples to help the reader fully understand the proofs.

This book hopes to bring those interested in Mathematics closer to the wonderful solutions that the best mathematicians in history have left us some centuries ago, so that they can enjoy them as I once did. These problems are somehow a compliment to the progress of Humanity.

Víctor Lauria

Contents

Chapter 1

Pythagorean theorem

(Pythagoras – 530 BC)

PROBLEM

To prove that in a right triangle the square of the hypotenuse is equal to the sum of the squares of the other sides.

HISTORY

Pythagoras (580 BC – 495 BC) was born on the island of Samos (Asia Minor) and he is considered to be the first pure mathematician in history. He was the son of a merchant that gave him a good education and the possibility of traveling around the known world at the time (Egypt, Arabia, Phenicia, Babylon, India), and in his youth he was banished from his homeland as a consequence of a war of which he became prisoner. Upon gaining freedom, he emigrated to Crotona (a city of "Magna Grecia" at that time and now part of the south of Italy), where he founded the school of astronomers, musicians, mathematicians and philosophers that would later be known as the Pythagorean school.

"The School of Athens" – Fresco of Raphael (1511)
Pythagoras is represented in the foreground, on the left

At that time, the authority of the teacher was sacred and every discovery of the school was attributed to him. Therefore, there are doubts as to whether the famous theorem that bears his name was really a finding of his own or some outstanding student's whose name will never be known. The Pythagorean school and its followers expanded rapidly after 500 BC, but from then on the school became politicized to such an extent that in 460 BC their members began to be fiercely attacked by their enemies until its disappearance.

SOLUTION

The exact method used by the Pythagoreans to prove this famous theorem is unknown, but it should certainly be based on the property that Thales of Miletus, who was a professor of Pythagoras, had found for similar triangles. Therefore, if we want to be rigorous we shall first prove some preliminary results that might have been always taken for granted.

First, let us see how to calculate the area of a triangle (as I said, we start with the very basics); in his masterpiece "Elements", Euclid gave such importance to this fact that it deserved a whole proposition (I.41).

LEMMA 1.1. *All triangles with equal base and equal height have the same area, which equals half the product of both quantities.*

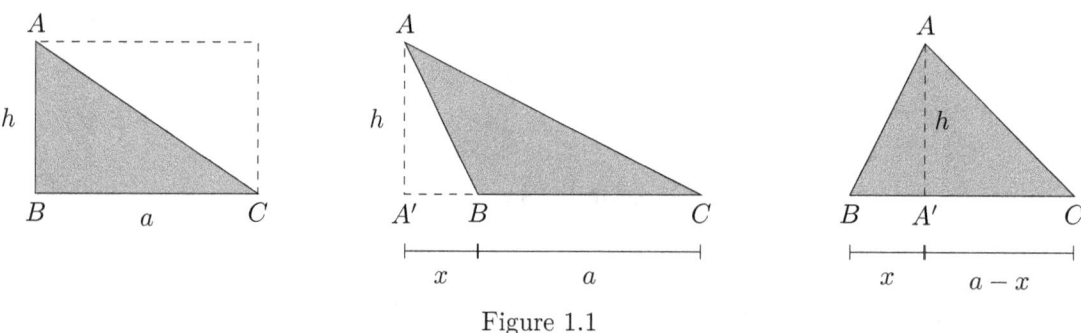

Figure 1.1

PROOF. It is clear that in a right triangle the lemma is true because its area is half the area of the rectangle that shares both sides with it (first graph of figure 1.1)

Consider now instead any triangle ABC where we take side BC as base (with length a), and suppose that its height (distance from vertex A to side BC) is h.

In the case that the projection of vertex A onto the line containing BC (point A') does not lie within segment BC (second graph of figure 1.1), the area of the triangle ABC can be calculated as the subtraction of the areas of the right triangles $AA'C$ and $AA'B$. So:

$$Area(ABC) = \tfrac{h \cdot (a+x)}{2} - \tfrac{h \cdot x}{2} = \tfrac{h \cdot a}{2}$$

In contrast, in the case that A' lies within segment BC (third graph of figure 1.1), the area of the triangle ABC can be calculated as the sum of the areas of the right triangles $AA'B$ and $AA'C$. So:

$$Area(ABC) = \tfrac{h \cdot x}{2} + \tfrac{h \cdot (a-x)}{2} = \tfrac{h \cdot a}{2}$$

In any case, the lemma is true. □

After this lemma we prove the so-called "first Thales Theorem", in honor of the first great scientist of the western civilization (Thales of Miletus, 624 BC – 546 BC). This is explained many times to elementary students at school as something "obvious" and it is never revisited again to prove it. The proof that we present only needs lemma 1.1.

THEOREM 1.1. *(Thales) Let r_1 and r_2 be two parallel lines, and let S be a point that is not between them nor belong to them. If we draw from S two lines that cut r_1 at points A and C, and r_2 at points B and D, (first graph of figure 1.2), then:*

$$\left|\frac{AC}{BD}\right| = \left|\frac{SA}{SB}\right| = \left|\frac{SC}{SD}\right|$$

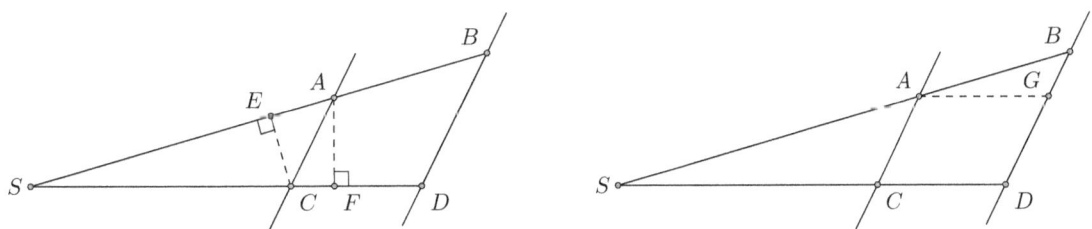

Figure 1.2

PROOF. Since r_1 and r_2 are parallel lines, triangles CDA and CBA have the same height (taking CA as base of both triangles). Then, by lemma 1.1, triangles CDA and CBA have the same area, which implies that triangles SCB and SDA also have the same area. From both statements we deduce that:

$$\frac{Area(SCA)}{Area(SDA)} = \frac{Area(SCA)}{Area(SCB)}$$

Let F be the projection of point A onto line SC and let E be the projection of the point C onto line SA. Then, applying lemma 1.1 again, the previous identity can be written as:

$$\frac{|SC|\cdot|AF|/2}{|SD|\cdot|AF|/2} = \frac{|SA|\cdot|EC|/2}{|SB|\cdot|EC|/2} \Rightarrow \frac{|SC|}{|SD|} = \frac{|SA|}{|SB|}$$

To prove the other equality, let us now set an additional line, parallel to SD and passing through A, which cuts BD at point G (second graph of figure 1.2). We can apply the same reasoning that we have used before, but now taking B as the outer point (it was S before) and AG, SD as parallel lines (they were AC and BD before), which leads us to:

$$\frac{|AB|}{|SB|} = \frac{|BG|}{|SB|} \Rightarrow \frac{|SB|-|SA|}{|SB|} = \frac{|BD|-|DG|}{|BD|} \Rightarrow \frac{|SA|}{|SB|} = \frac{|DG|}{|BD|} \Rightarrow \frac{|SA|}{|SB|} = \frac{|AC|}{|BD|}$$

$\square$

This theorem leads us directly to a corollary that is the property of "similarity of triangles", also widely known.

COROLLARY 1.1. *If we have two similar triangles ABC and $A'B'C'$ (that is, the angle of vertex A [resp. B, C] is equal to the angle of vertex A' [resp. B', C']), its sides are in proportion:*

$$\frac{|AB|}{|A'B'|} = \frac{|AC|}{|A'C'|} = \frac{|BC|}{|B'C'|}$$

PROOF. Two similar triangles can be drawn as triangles SAC and SBD in figure 1.2, since their angles are equal. It is achieved, for example, by matching vertices A and A', and by drawing vertices B and B' [resp. C and C'] in the same line starting from A (see figure 1.3).

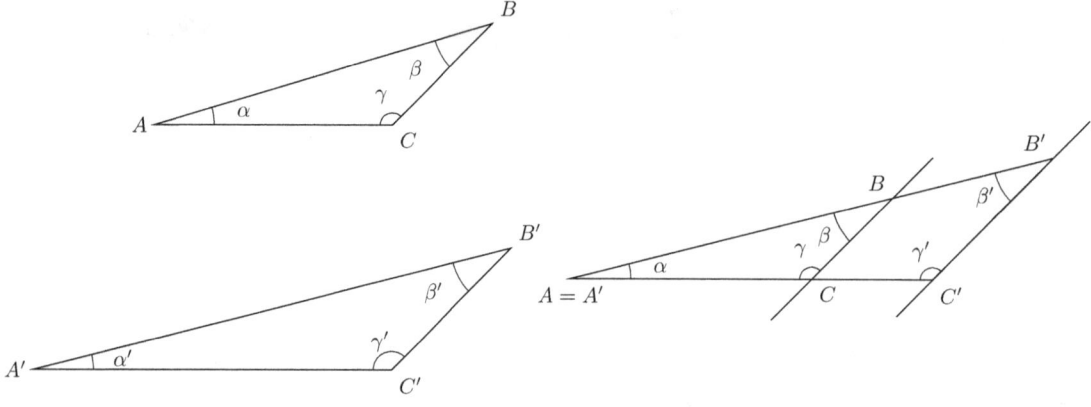

Figure 1.3

Then, we can apply Thales's theorem to find the equalities we want to prove (proportion between sides). □

We are now in position to see what a member of the Pythagorean school found 2500 years ago.

THEOREM 1.2. *(Pythagoras) In a right triangle, the square of the hypotenuse is equal to the sum of the squares of the other sides.*

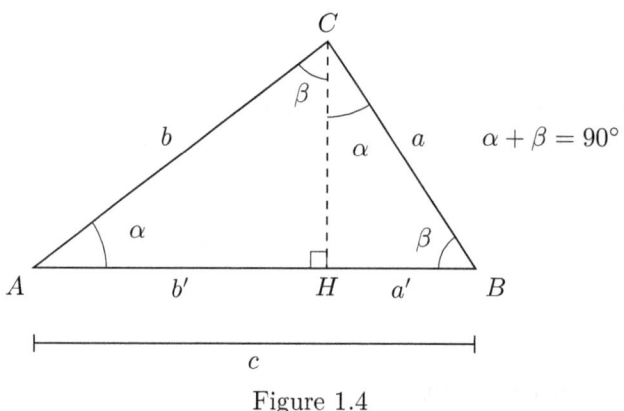

$$\alpha + \beta = 90°$$

Figure 1.4

PROOF. In figure 1.4, ACB is a right triangle (right angle at vertex C); we want to prove that its sides fulfill the property $a^2 + b^2 = c^2$.

Let point H be the projection of vertex C onto side AB, splitting it into two parts: AH (length b') and BH (length a').

Then, as ACB is a right triangle, it is satisfied that ACB, AHC and CHB are similar triangles, since all of them have a right angle and, in addition, the angle in vertex A of ACB is equal to the angle in vertex A of AHC (obvious) and equal to the angle in vertex C of CHB (because this angle is the complementary of the angle in vertex C of AHC).

Now, as ACB and AHC are similar triangles, we conclude by applying the corollary for similar triangles of Thales's theorem that:

$$\frac{c}{b} = \frac{b}{b'}$$

Similarly, but now because ACB and CHB are similar triangles:

$$\frac{c}{a} = \frac{a}{a'}$$

Both equations can be written as $b^2 = c \cdot b'$ and $a^2 = c \cdot a'$, which results, if we add them, in $a^2 + b^2 = c \cdot (a' + b') = c^2$. ☐

As I said before, it is believed that this was the proof that the Pythagoreans once found, but many other proofs have been discovered later on; the interested reader may find a large number of them on his own. Nevertheless, few will surpass in beauty Euclid's' proof, published in his work "Elements"; I have not resisted the temptation to include it in this book.

THEOREM 1.3. *(Euclid) In a right triangle, the square of the hypotenuse is equal to the sum of the squares of the other sides.*

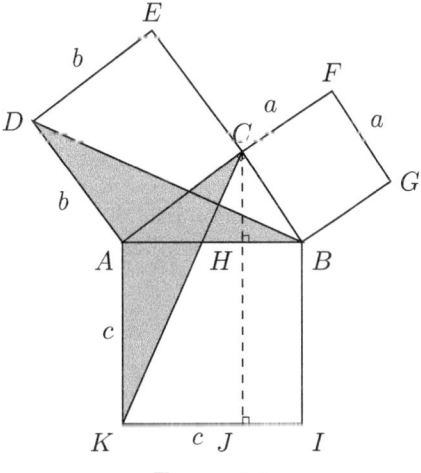

Figure 1.5

PROOF. In figure 1.5 we have drawn three squares such that each of them has a side that equals one of the sides of the original (right) triangle. We want to prove that the area of the largest square (of side equal to $AB = c$) is equal to the sum of the areas of the other two squares, that is, $a^2 + b^2 = c^2$.

Triangle CAK is identical to triangle DAB (both have have equal sides of length b and c, and both have an equal angle in vertex A, which has the value of a right angle plus the magnitude of the angle of vertex A in the original triangle), so their areas are equal.

Furthermore, by lemma 1.1, the area of triangle CAK is half the area of rectangle $AHJK$, while the area of triangle DAB is half the area of square $ACED$ (because the angle in vertex C of triangle ABC is a **right** angle; otherwise, the statement would be false). Therefore, the areas of rectangle $AHJK$ and square $ACED$ are equal.

By an analogous reasoning, the areas of rectangle $BHJI$ and square $BCFG$ are also equal. Bringing both conclusions together, the sum of the areas of squares $ACED$ and $BCFG$ is equal to the sum of the areas of rectangles $AHJK$ and $BHJI$, i.e., the area of square $ABIK$. Therefore, $a^2 + b^2 = c^2$. ☐

FINAL REMARKS

The discovery of the theorem is supposed to have brought great joy to the Pythagorean school. However, shortly after that they discovered a consequence of the theorem that left them terrified, to the point that the brotherhood forbade making it public: the irrationality of $\sqrt{2}$.

Indeed, the Pythagoreans were defenders of numbers and order, and for them the numbers "ended" in the rational numbers. Upon discovering the theorem, they realized that a right triangle with legs of length equal to 1 has a hypotenuse of a length that equals a number whose square is 2 ($x = \sqrt{2}$ in current notation).

Studying this number, they proved that it could not be a rational number (that is, a quotient of integers). To see this, let us suppose that x is a rational number and so that it can be written as a quotient of two (positive) integers: $\sqrt{2} = p/q$. We can suppose that p and q do not share a common divisor, since in that case we can divide the fraction by that divisor until finding another ($\sqrt{2} = p'/q'$) that no longer has one.

Then, we have $\sqrt{2} = p/q \Rightarrow 2 = p^2/q^2$, which implies that $2q^2 = p^2$ and, therefore, that p must be even. If p is even, we can write it as $p = 2k$ (where k is an integer) and now we have $2q^2 = 4k^2 \Rightarrow q^2 = 2k^2$, which now implies that q must also be even. But this is a contradiction, since if both p and q are even, they have 2 as a common divisor, which contradicts our assumption. The hypothesis that $\sqrt{2}$ is a rational number must be false.

As we have already said, the Pythagoreans did not like this fact and decided to keep it secret; unconfirmed legends even claim that they killed a member of the school who intended to reveal it.

Chapter 2

Doubling the cube

(Menaechmus – 335 BC)

PROBLEM

Given the edge of a cube, to find geometrically the edge of a second cube whose volume is double that of the first.

HISTORY

In 429 BC an outbreak of plague ravaged Greece, and more especially the city of Athens, whose governor Pericles was one of the most illustrious victims. According to the legend, the Oracle of Delos (don't confuse with the Oracle of Delphi, much more famous but not a part of this story), one of the Greek islands of the Cyclades, was consulted about how to stop the epidemic. The answer of the Oracle (which was capricious as usual) was that the plague would be overcome if the inhabitants of Athens built a new cubical altar whose volume was exactly the double of the existing one.

Ruins of Delos (Greece)
Photograph: Ggia (Creative Commons)

The veracity of this legend is difficult to believe, but the truth is that the problem was posed in those terms and important mathematicians of antiquity tried to solve it. As we will see in the final conclusions, the problem cannot be solved with the help of a ruler and compass alone (hence its fame), but that could not be proven until some centuries later.

The Greek mathematician Menaechmus, who was one of the tutors of Alexander the Great (as was Aristotle), found a solution based on parabolas in 335 BC and that's why his name is associated with the problem. However, here we will present a later method, based on the so-called "construction with paper strip", a solution that does not use conics (which, anyway, cannot be drawn exactly) and it is "close" to a ruler-and-compass construction, since the only auxiliary element used is a strip of paper where a segment of a certain length has previously been drawn.

SOLUTION

Let k be the edge of a cubic altar, so its volume is k^3. If we want to double the volume of this altar keeping its cubic shape, we must construct a segment of length $\sqrt[3]{2k^3} = \sqrt[3]{2} \cdot k$. The problem is to do it with an **exact** method and using geometry: that is, we must start from a segment of length k and, constructing parallels, perpendiculars, circumferences, angles, etc., obtain another segment of length $\sqrt[3]{2} \cdot k$.

To give a much simpler example, imagine that we would like to obtain a segment of length $\sqrt{5} \cdot k$. In that case, it would only be necessary to draw a segment of length k, draw another one of length $2k$ which is perpendicular to the first one and starts in one of its ends and then complete a triangle with another segment. By the Pythagorean theorem, the hypotenuse of the triangle has a length of $\sqrt{k^2 + (2k)^2} = \sqrt{5} \cdot k$, which would be an exact solution to the problem.

But if we look for a segment of length $\sqrt[3]{2} \cdot k$, then the process is not so easy. To understand the solution that will be proposed later, it is first convenient to study two theorems of classical geometry.

THEOREM 2.1. *(Transverse line theorem) Let ABC be a triangle and r a line (called transverse) that cuts two sides of it (for example, sides AB and AC) and the line containing side BC. Let D, E and F be, respectively, the three points of intersection (see figure 2.1). Then:*

$$\overline{AD} \cdot \overline{BF} \cdot \overline{CE} = \overline{AE} \cdot \overline{BD} \cdot \overline{CF}$$

PROOF. Draw the three lines perpendicular to the direction of the side BC and passing through points A, B and C respectively. Let A', B' and C' be the intersection points of these lines with line r, as we see in figure 2.1. Let a, b and c be the distances $a = \overline{AA'}$, $b = \overline{BB'}$ and $c = \overline{CC'}$.

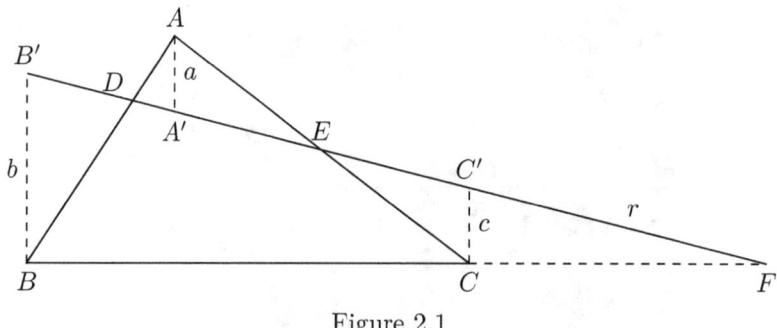

Figure 2.1

$BB'F$ and $CC'F$ are similar triangles, so $\overline{BF}/\overline{CF} = b/c$; similarly, $BB'D$ and $AA'D$ are similar triangles, so $\overline{AD}/\overline{BD} = a/b$; and finally, $AA'E$ and $CC'E$ are similar triangles, so $\overline{CE}/\overline{AE} = c/a$. Multiplying the three equations, we can eliminate a, b and c:

$$\frac{\overline{AD}}{\overline{BD}} \cdot \frac{\overline{BF}}{\overline{CF}} \cdot \frac{\overline{CE}}{\overline{AE}} = 1 \Rightarrow \overline{AD} \cdot \overline{BF} \cdot \overline{CE} = \overline{AE} \cdot \overline{BD} \cdot \overline{CF}$$

$\square$

THEOREM 2.2. *(Stewart) Let ABC be a triangle and let d be the length of a segment that joins vertex A with a point P of side BC. Let a, b and c be the lengths of sides BC, AC and AB, respectively, and let m and n be the lengths of the segments into which side BC is split by point P, so $m = \overline{BP}$ and $n = \overline{CP}$ (see figure 2.2). Then:*

$$b^2m + c^2n = a \cdot (d^2 + mn)$$

PROOF. Let A' be the projection of vertex A onto side BC and suppose that A' lies between points B and P (the reasoning would be analogous if it lies between points C and P, or even if A' coincides with point P). Let x be the distance BA' and, therefore, let $m - x$ be the distance $A'P$.

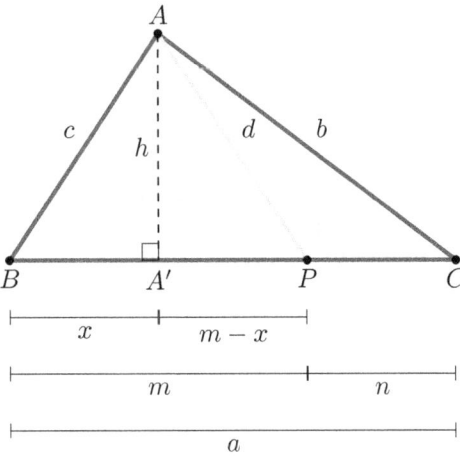

Figure 2.2

Looking at the right triangles of figure 2.2 ($AA'B$, $AA'P$ and $AA'C$) we deduce, by the Pythagorean theorem, the following three equations:

(2.1)
$$\begin{cases} c^2 = h^2 + x^2 \\ d^2 = h^2 + (m - x)^2 \\ b^2 = h^2 + (n + m - x)^2 \end{cases}$$

Now, we use the first equation of (2.1) to substitute in the last two, eliminating variable h:

(2.2)
$$\begin{cases} d^2 = c^2 + m^2 - 2mx \\ b^2 = c^2 + n^2 + m^2 + 2mn - 2nx - 2mx \end{cases}$$

And now we use the first equation of (2.2) in the second, eliminating variable x:

$$b^2 = d^2 + n^2 + 2mn - 2nx \Rightarrow b^2 = d^2 + n^2 + 2mn - 2n \cdot \left(\frac{c^2 + m^2 - d^2}{2m} \right) \Rightarrow$$

$$\Rightarrow b^2m = d^2m + n^2m + 2nm^2 - n \cdot (c^2 + m^2 - d^2) \Rightarrow$$

$$\Rightarrow b^2m + c^2n = d^2 \cdot (m + n) + m^2n + n^2m \Rightarrow$$

$$\Rightarrow b^2m + c^2n = d^2 \cdot (m + n) + mn \cdot (m + n) = (m + n) \cdot (d^2 + mn) = a \cdot (d^2 + mn)$$

$\square$

Now we are ready to present the method to draw a segment of length $\sqrt[3]{2} \cdot k$ and to understand the proof.

Construction of a segment of length $\sqrt[3]{2} \cdot k$ starting with a segment of length k

Steps for the construction:

(1) Draw an equilateral triangle ABC with sides of length k (possible with a compass)
(2) Extend side AC (starting at vertex A) with an additional length k (possible with ruler-and-compass) to get point D
(3) Join points D and B by sufficiently extending the line beyond the point B (possible with a ruler)
(4) Extend sufficiently side AB beyond point B (possible with a ruler)
(5) Place conveniently the strip of paper (where a segment of length k has been drawn before) until an end point of the segment lies on the extension of side AB, the other end point lies on the extension of side BD and the edge of the paper passes through point C, as shown in figure 2.3 (this step is the only one that **CANNOT** be done with ruler-and-compass)
(6) Let P be the end point of the segment that lies on the extension of side AB and let Q be the end point of the segment that lies on the extension of side BD. Then, the distance between Q and C is $\sqrt[3]{2} \cdot k$

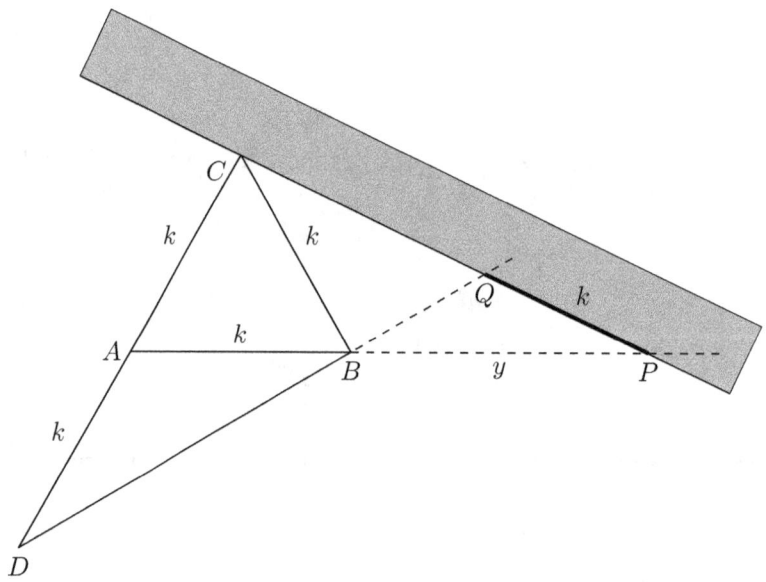

Figure 2.3

Proof that $\overline{CQ} = \sqrt[3]{2} \cdot k$

Consider, in figure 2.3, the triangle ACP and the transverse line DBQ. Let x be the distance CQ (we want to see that it is equal to $\sqrt[3]{2} \cdot k$, but we don't know that yet) and let y be the distance BP (unknown). Then, by Theorem 2.1:

(2.3) $$\overline{AD} \cdot \overline{CQ} \cdot \overline{BP} = \overline{AB} \cdot \overline{PQ} \cdot \overline{CD} \Rightarrow k \cdot x \cdot y = k \cdot k \cdot 2k \Rightarrow xy = 2k^2$$

On the other hand, if we consider triangle ACP and segment CB, then, by Theorem 2.2:

$$\overline{AC}^2 \cdot \overline{BP} + \overline{CP}^2 \cdot \overline{AB} = \overline{AP} \cdot (\overline{CB}^2 + \overline{AB} \cdot \overline{BP}) \Rightarrow k^2 \cdot y + (x+k)^2 \cdot k = (k+y) \cdot (k^2 + k \cdot y) \Rightarrow$$

(2.4) $$\Rightarrow k \cdot y + (x + k)^2 = (k + y) \cdot (k + y) \Rightarrow x^2 + 2xk = ky + y^2$$

If we isolate the variable y from equation (2.3) and substitute its value in (2.4):

$$x^2 + 2xk = \frac{2k^3}{x} + \frac{4k^4}{x^2} \Rightarrow x^4 + 2x^3 k - 2xk^3 - 4k^4 = 0$$

Finally, we apply $x = r \cdot k$ (everything is in proportion to the length k given at the beginning) and the above equation becomes $r^4 + 2r^3 - 2r - 4 = 0$, where we can check that $r = \sqrt[3]{2}$ is one of its solutions. As we see in figure 2.4, the equation has only one more real solution, but it is a negative one. Therefore, it follows that $x = \overline{CQ} = \sqrt[3]{2} \cdot k$ and we have managed to "double the cube".

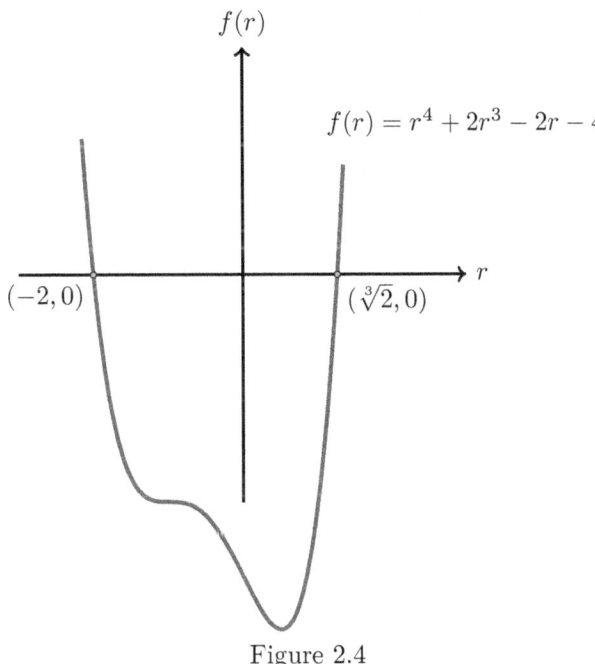

Figure 2.4

FINAL REMARKS

To show that a segment of length $\sqrt[3]{2}$ cannot be constructed with ruler-and-compass (from another segment whose length is a rational number) it would be necessary to study non-elementary mathematics. However, with the following example I hope to "convince" the reader.

Let us see at figure 2.5, where we have started with a segment of length 1 (in fact, we could have done it from any rational number) and we have placed it in such a way that its end points are $O = (0,0)$ and $A = (1,0)$. Next, we have constructed point $B = (\sqrt{5}, 0)$ with a segment of length 2 that is perpendicular to axis X and applying the Pythagorean Theorem (all steps can be easily done with ruler-and-compass).

As a second step, we subtract a unit length to point B on axis X (obtaining $C = (\sqrt{5} - 1, 0)$) and we split segment OC into four parts to take one of them (part next to the origin). Let us denote $D = ((\sqrt{5} - 1)/4, 0)$ to the end point found. Starting at point D we build a perpendicular line to axis X and let us set $E = ((\sqrt{5} - 1)/4, (\sqrt{10 + 2\sqrt{5}}))$ to the intersection of this line and the circumference of center O and radius equal to 1.

11

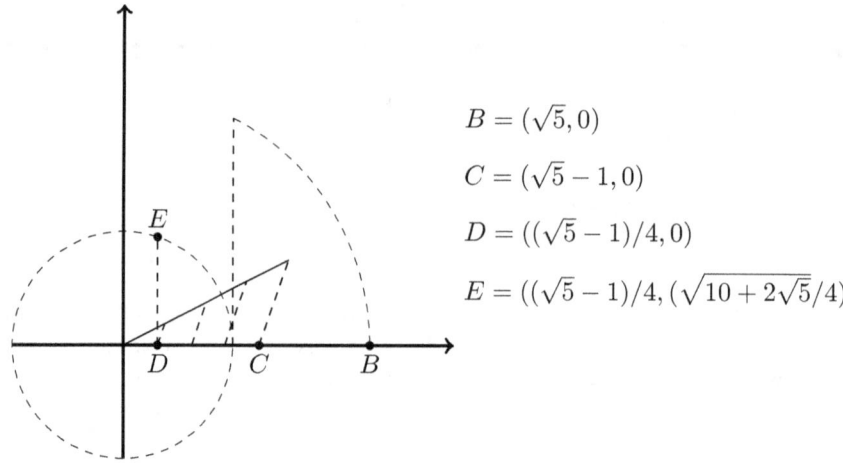

$$B = (\sqrt{5}, 0)$$

$$C = (\sqrt{5} - 1, 0)$$

$$D = ((\sqrt{5} - 1)/4, 0)$$

$$E = ((\sqrt{5} - 1)/4, (\sqrt{10 + 2\sqrt{5}}/4))$$

Figure 2.5

Continuing in this way (drawing circles, parallel or perpendicular lines, lines passing through two known points, looking for intersections with circles and previous lines and so on) we are doing the so-called "ruler-and-compass constructions". It can be proved that the coordinates of points found with these steps fulfill the property of being solutions of polynomials of power-of-two degree with rational coefficients and irreducible over the rational numbers.

In our example:

- $B = (\sqrt{5}, 0)$ satisfies that its first coordinate is solution of the polynomial $p(x) = x^2 - 5$ (polynomial of power-of-two degree with rational coefficients and irreducible over the rational numbers).
- $C = (\sqrt{5} - 1, 0)$ satisfies that its first coordinate is solution of the polynomial $p(x) = x^2 + 2x - 4$ (polynomial of power-of-two degree with rational coefficients and irreducible over the rational numbers).
- $D = ((\sqrt{5} - 1)/4, 0)$ satisfies that its first coordinate is solution of the polynomial $p(x) = x^2 + x - 1$ (polynomial of power-of-two degree with rational coefficients and irreducible over the rational numbers).
- Finally, $E = ((\sqrt{5} - 1)/4, (\sqrt{10 + 2\sqrt{5}}))$ has two nonzero coordinates: the first one has already been studied, while the second is solution of the polynomial $p(x) = 16x^4 - 20x^2 + 5$ (polynomial of power-of-two degree with rational coefficients and irreducible over the rational numbers).

We can continue drawing points with ruler-and-compass steps, and later check this property. Obviously, the more complicated the construction, the greater degree of the polynomial and the more difficult it will be to prove its irreducibility, but the condition will continue to be satisfied.

Now we just need to see what happens with number $\sqrt[3]{2}$. The property that we will use, ignoring its proof, is the following: "*If a real number is a solution of a polynomial with rational coefficients and irreducible over the rational numbers, then it cannot be the solution of another polynomial with rational coefficients, irreducible over the rational numbers and different degree*".

In our case, $\sqrt[3]{2}$ is the solution of the 3-degree polynomial with rational coefficients and irreducible over the rational numbers $p(x) = x^3 - 2$. Therefore, by the previous property, it cannot be a solution of a polynomial with rational coefficients that is irreducible over the rational numbers and has a degree other than three (that includes, of course, a power-of-two degree polynomial). Therefore, a point with a coordinate equal to $\sqrt[3]{2}$ cannot be constructed with ruler-and-compass steps.

Chapter 3

Trisection of an angle

(Pappus of Alexandria – 300 BC)

PROBLEM

To divide geometrically any angle into three equal parts.

HISTORY

The trisection of the angle is not as well known as the problem of doubling the cube that we saw in the previous chapter, but it is also one of the great mathematical challenges of antiquity. It is not known when it was raised for the first time, but at least we know that Hippocrates of Chios (470 – 410 BC) was one of the first mathematicians to study it.

"Mathematicae Collectiones" – Pappus of Alexandria
Latin translation (1660)

The problem is to find a geometric method in order to divide any angle into three equal parts; it is evident that it can be easily done in concrete cases, as in the case of a 90° angle, but the Greeks wanted to know a way that worked for any angle. A solution was found by Pappus of Alexandria towards 300 BC using a hyperbola, as he described in his book "Mathematical Collections". However, we will see here a simpler method, discovered by Archimedes, that uses (as in the problem of doubling the cube) a strip of paper with a marked distance.

We will also see that the trisection of the angle cannot be done only with ruler-and-compass steps, which is the reason why it continued to intrigue illustrious mathematicians for centuries.

SOLUTION

a) Method to trisect any given angle

Steps to trisect angle ϕ with vertex in S:

- With center in S, draw a circle of any radius (let r be the length of this radius) that cuts the angle at points A and B (see figure 3.1).
- Conveniently place a strip of paper where we have previously drawn a segment of length r so that the strip passes through point B, one end point of the segment coincides with a point on the circumference (let P be this point) and that the other end point lies in the prolongation of AS outside the circumference (let Q be this point). This step CANNOT be done with ruler-and-compass.
- Angle $\alpha = \widehat{PQS}$ is the third part of the initial angle ϕ

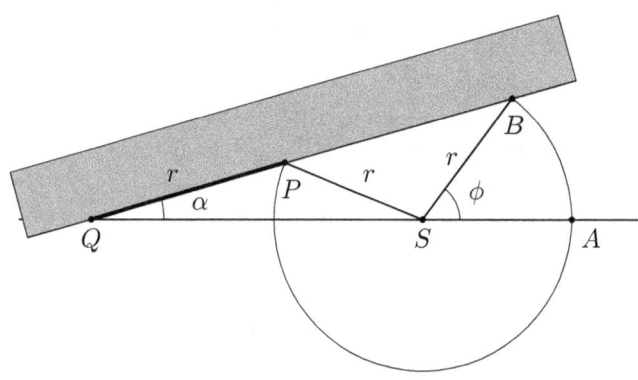

Figure 3.1

Proof:

It is clear that $\overline{PS} = \overline{PQ} = r$, so triangle PQS is isosceles, and then it follows that angles $\widehat{PQS}$ and $\widehat{PSQ}$ are equal (and equal to α, by definition of α). Therefore, angle $\widehat{QPS}$ is equal to $180 - 2\alpha$, and we can deduce that angle $\widehat{SPB}$ is equal to 2α.

Let us have a look now at triangle PBS: it is isosceles, since sides $\overline{SP}$ and $\overline{SB}$ are equal (both have length r), so angle $\widehat{PBS}$ is equal to the value 2α too. Therefore, if we let β be the angle $\widehat{PSB}$, from the sum of the three angles of that triangle (which must be equal to $180°$) we can deduce that:

$$2\alpha + 2\alpha + \beta = 180 \Rightarrow \beta = 180 - 4\alpha$$

Finally, we know that the sum of the three concurrent angles in S ($\widehat{QSP}$, $\widehat{PSB}$ and $\widehat{BSA}$) must be $180°$, so:

$$\alpha + \beta + \phi = 180 \Rightarrow \alpha + (180 - 4\alpha) + \phi = 180 \Rightarrow \alpha = \phi/3$$

b) Schonemann irreducibility criterion

Let us make a small parenthesis in our study of the trisection of an angle to get into the proof of a criterion to decide if a polynomial of integer coefficients is irreducible over the rational numbers.

14

DEFINITION 3.1. *A polynomial of integer coefficients is called a **primitive polynomial** if there is no prime number p that divides all its coefficients simultaneously (for example, $3x^3 + 12x^2 + 21$ is not primitive because prime $p = 3$ divides all the coefficients).*

LEMMA 3.1. *Multiplication of two primitive polynomials $f(x)$ and $g(x)$ is also a primitive polynomial.*

PROOF. Suppose that polynomial $h(x) = f(x) \cdot g(x)$ is not primitive. That means there would be a prime p that divides all coefficients of $h(x)$ but does not divide all coefficients of $f(x)$ nor all coefficients of $g(x)$:

$$h(x) = (a_m x^m + \cdots + a_i x^i + \cdots + a_0) \cdot (b_n x^n + \cdots + b_j x^j + \cdots + b_0)$$

Since p does not divide all coefficients of $f(x)$ at the same time, there must be one or more coefficients that are not multiples of p: let us choose the coefficient of **highest** degree, and let us suppose it is a_i. In a similar way, since p does not divide all coefficients of $g(x)$ at the same time, there is one or more coefficients that are not multiples of p: let us choose the coefficient of **highest** degree, and let us suppose it is b_j.

In that case, when multiplying both polynomials, the coefficient of degree x^{i+j} turns out to be the sum of $a_i \cdot b_j$ (which is **NOT** a multiple of p) and other addends of type $a_r \cdot b_s$ ($r + s = i + j$) where either $r > i$ or $s > j$. In any case, either a_r is a multiple of p (if $r > i$, because a_i is the multiple of p of highest degree) or b_s is a multiple of p (if $s > j$, for equivalent reason). Therefore, the coefficient of x^{i+j} is a sum of a term that is NOT multiple of p and an indeterminate number of terms that ARE multiple of p, which leads to the conclusion that the sum is NOT a multiple of p.

But this is a contradiction, since we have assumed that p divides all coefficients of $h(x)$. The initial hypothesis ($h(x)$ is not primitive) was wrong. $\square$

DEFINITION 3.2. *Let $f(x)$ be a polynomial with integer coefficients. The **content of the polynomial** f, $\mathrm{cont}(f)$, is the greatest common divisor of its coefficients (for example, content of $f(x) = 3x^3 + 12x^2 + 21$ is 3, that is, $\mathrm{cont}(f) = 3$). It is clear that the property of being a primitive polynomial is the same as the property of a polynomial with content equal to 1.*

*The definition can be extended to polynomials of rational coefficients in the following way: let $f(x)$ be a polynomial of rational coefficients and let n be an integer such that $n \cdot f(x)$ is a polynomial of integer coefficients. We call **content of the polynomial** f and we write $\mathrm{cont}(f)$ to the number $\mathrm{cont}(n \cdot f)/n$. For example, to calculate the content of $f(x) = (2/3)x^2 + 5x + 1/3$, we multiply it by $n = 3$ to get a polynomial of integer coefficients, $3 \cdot f(x) = 2x^2 + 15x + 1$; then, the content of the original polynomial is the division of the content of this new polynomial (which is 1) by 3, that is, $\mathrm{cont}(f) = \mathrm{cont}(3 \cdot f)/3 = 1/3$.*

It should be noted that if we choose another n that also satisfies that $(n \cdot f)$ is a polynomial of integer coefficients, then the content will be the same. For example, by choosing $n = 9$ we have $9 \cdot f(x) = 6x^2 + 45x + 3$ (whose content is 3) and, therefore, $\mathrm{cont}(f) = \mathrm{cont}(9 \cdot f)/9 = 3/9 = 1/3$.

It should also be noted that if by chance f turns out to be a polynomial with integer coefficients, this definition matches the one we gave for them (since we apply the formula for $n = 1$).

LEMMA 3.2. *Only polynomials with integer coefficients have a content that is equal to an integer.*

PROOF. Let $f(x) = a_d x^d + \cdots + a_1 x + a_0$ be a polynomial of rational coefficients whose content is an integer. We want to see that necessarily all its coefficients are integers.

Let m be the minimum integer such that $m \cdot f(x) = m \cdot a_d x^d + \cdots + m \cdot a_1 x + m \cdot a_0$ has all integer coefficients. By hypothesis, the content of f is an integer:

$$\text{cont}(f) = \frac{1}{m} \cdot \gcd(m \cdot a_d, \cdots, m \cdot a_1, m \cdot a_0)$$

which means that $\gcd(m \cdot a_d, \cdots, m \cdot a_1, m \cdot a_0)$ is an integer that is multiple of m. This implies that all $m \cdot a_i$ are multiples of m, i.e., all a_i are integers. $\square$

LEMMA 3.3. *If $f(x)$ is a polynomial with rational coefficients and q is a rational number, then $cont(q \cdot f) = q \cdot cont(f)$*

For example, we saw earlier that the content of $f(x) = (2/3)x^2 + 5x + 1/3$ was $1/3$. If we multiply the polynomial by $5/2$ we obtain $5/2 \cdot f(x) = (5/3)x^2 + (25/2)x + 5/6$, whose content is $1/3 \cdot 5/2 = 5/6$.

PROOF. Let $q = a/b$, where a and b are integers, and let us suppose that n is an integer such that $n \cdot f(x)$ is a polynomial of integer coefficients. That means that $an \cdot f(x)$ is also a polynomial with integer coefficients and, as $a = b \cdot q$, then $bqn \cdot f(x)$ is also a polynomial with integer coefficients too.

Therefore, we have the equations:

(3.1) $$bn \cdot \text{cont}(q \cdot f) = \text{cont}(bnq \cdot f) = \text{cont}(an \cdot f) = a \cdot \text{cont}(n \cdot f)$$

where the first and the third equalities are deduced from the definition of the content of a polynomial with integer coefficients (we are multiplying the coefficients by an integer and therefore the greatest common divisor must also be multiplied by that integer).

Finally, we obtain:

$$\text{cont}(q \cdot f) = \frac{1}{bn} \cdot \text{cont}(bnq \cdot f) = \frac{1}{bn}\text{cont}(an \cdot f) = \frac{a}{bn} \cdot \text{cont}(n \cdot f) = \frac{a}{b} \cdot \text{cont}(f) = q \cdot \text{cont}(f)$$

where the first and the third equalities are deduced from (3.1), and the fourth equality is the application of the definition of the content of a polynomial with rational coefficients. $\square$

LEMMA 3.4. *Let $f(x)$ and $g(x)$ be two polynomials with rational coefficients. Then, $cont(f \cdot g) = cont(f) \cdot cont(g)$*

PROOF. We define the new polynomials:

$$f_1(x) = \frac{f(x)}{\text{cont}(f)} \qquad g_1(x) = \frac{g(x)}{\text{cont}(g)}$$

that is, we divide the coefficients by their greatest common divisor. It is not strange to verify that we get that the content of both is 1 (equivalently, that they are primitive polynomials):

$$\text{cont}(f_1) = \text{cont}\left(\frac{1}{\text{cont}(f)} \cdot f\right) = \frac{1}{\text{cont}(f)} \cdot \text{cont}(f) = 1$$

where in the second equality we have applied lemma 3.3.

Now, applying lemma 3.1, we deduce that polynomial $f_1 \cdot g_1$ is primitive (that is, its content is 1) and from there we infer the final conclusion:

$$\text{cont}(f \cdot g) = \text{cont}(\text{cont}(f) \cdot \text{cont}(g) \cdot f_1 \cdot g_1) =$$
$$= \text{cont}(f) \cdot \text{cont}(g) \cdot \text{cont}(f_1 \cdot g_1) = \text{cont}(f) \cdot \text{cont}(g)$$

where in the last equality we have applied what we have just seen ($f_1 \cdot g_1$ is primitive). $\qquad\square$

LEMMA 3.5. *(Gauss) If a primitive polynomial with integer coefficients $h(x)$ is divisible over the rational numbers (i.e., it is the product of two nontrivial polynomials $f(x)$ and $g(x)$ with rational coefficients), then there is also another decomposition of $h(x)$ as a product of two non-trivial polynomials with **integer** coefficients.*

That is, if a primitive polynomial with integer coefficients is reducible over the rational numbers, it is also reducible over the integers. Let us see the proof.

PROOF. Set $h(x) = f(x) \cdot g(x)$ under the conditions of the hypothesis. We define again the new polynomials:

$$f_1(x) = \frac{f(x)}{\text{cont}(f)} \qquad\qquad g_1(x) = \frac{g(x)}{\text{cont}(g)}$$

and, as we have just seen in the previous lemma, the content of both is 1 (primitive polynomials).

Now:

$$f_1(x) \cdot g_1(x) = \frac{f(x) \cdot g(x)}{\text{cont}(f) \cdot \text{cont}(g)} = \frac{h(x)}{\text{cont}(f) \cdot \text{cont}(g)}$$

which implies that:

$$\text{cont}(f_1 \cdot g_1) = \frac{1}{\text{cont}(f) \cdot \text{cont}(g)} \cdot \text{cont}(h) \qquad\Rightarrow$$

$$\text{cont}(f_1) \cdot \text{cont}(g_1) = \frac{1}{\text{cont}(f) \cdot \text{cont}(g)} \cdot \text{cont}(h) \qquad\Rightarrow$$

$$1 = \frac{1}{\text{cont}(f) \cdot \text{cont}(g)} \cdot \text{cont}(h)$$

where we have applied lemma 3.3 in the equality of the first line, lemma 3.4 in the step from the first to the second line, and the fact that $f_1(x)$ and $g_1(x)$ are primitive in the step from the second to the third line.

$h(x)$ is primitive by hypothesis (its content is 1) so the last equality implies that cont$(f)\cdot$cont$(g)=1$, which leads us to:

$$f_1(x) \cdot g_1(x) = \frac{f(x) \cdot g(x)}{\text{cont}(f) \cdot \text{cont}(g)} = f(x) \cdot g(x)$$

We have started from a decomposition of $h(x)$ over the rational numbers $(h(x) = f(x) \cdot g(x))$ and we have finished with another decomposition over the integers $(h(x) = f_1(x) \cdot g_1(x))$. $\square$

THEOREM 3.1. *(Schonemann's irreducibility criterion) Let $f(x) = C_n x^n + C_{n-1} x^{n-1} + \cdots + C_1 x + C_0$ be a primitive polynomial with integer coefficients that satisfies the property that all its coefficients except C_n are divisible by a prime number p, and that C_0 is not divisible by p^2. Then, $f(x)$ is irreducible over the rational numbers.*

PROOF. Suppose that $f(x)$ is primitive and reducible over the rational numbers, and that will lead us to a contradiction.

By Gauss's lemma, $f(x)$ is reducible over the integers and therefore we can write $f(x) = (A_m x^m + \cdots + A_0) \cdot (B_k x^k + \cdots + B_0)$, where all the coefficients A_i, B_j are integers and where $m + k = n$.

Comparing coefficients C_i with the way they are calculated by multiplying the coefficient polynomials A_i, B_j, we have the following set of equations (note that we only write the m first equations, when in fact there are n equations):

$$\begin{cases} C_0 = A_0 \cdot B_0 \\ C_1 = A_0 \cdot B_1 + A_1 \cdot B_0 \\ C_2 = A_0 \cdot B_2 + A_1 \cdot B_1 + A_2 \cdot B_0 \\ \cdots \\ C_m = A_0 \cdot B_m + \cdots + A_m \cdot B_0 \end{cases}$$

C_0 is divisible by p but not by p^2 by hypothesis, then the first equation tells us that either A_0 or B_0 (but not both) is a multiple of p; suppose it is A_0 (the reasoning would be similar in the case that we chose B_0, but then we would use the first k equations to calculate coefficients C_i instead of the first m ones).

This implies, in the second equation, that A_1 must be a multiple of p (since both C_1, by hypothesis, and A_0, by the previous paragraph, are multiples of p, and B_0 is not). The reasoning is similar as we move through i-th equation: C_i are multiples of p, $A_0, \cdots, A_{i-1}$ are multiples of p too by the previous equations, and B_0 is not, so that's why it is necessary that A_i is a multiple of p.

The reasoning ends in the m-th equation, with the need for A_m to be a multiple of p. That is, all coefficients A_i must be multiples of p and, as we have $f(x) = (A_m x^m + \cdots + A_0) \cdot (B_k x^k + \cdots + B_0)$, then all C_i must be multiples of p too, including C_n. But this is a contradiction, since the hypothesis of the statement specified that C_n **NOT** a multiple of p. We have found a contradiction, so the assumption that $f(x)$ is reducible over the rational numbers was wrong. $\square$

Schonemann's criterion is not a perfect tool to decide whether a polynomial is irreducible over the rational numbers or not, since it does not say anything about the irreducibility of polynomials that do **not** meet the conditions. For example, the criterion says that $f(x) = 4x^3 + 7x^2 + 21x - 14$ is irreducible over the rational numbers (because there is a prime, 7, that divides all the coefficients except the one with highest degree, and because $p^2 = 49$ does not divide C_0), but it says nothing about $f(x) = 4x^3 + 7x^2 + 21x - 15$, for example, since there is no p that meets the required specifications.

c) Impossibility to trisect all angles with ruler-and-compass

Using Schonemann's criterion and the argument explained in the problem of doubling the cube, we will now prove that there is no alternative method that leads to the solution of the trisection of any angle with ruler-and-compass steps.

If we draw angle ϕ (that we want to trisect) so its vertex is in the origin of coordinates, and axis X corresponds to one of its sides, we can draw the circumference of radius 1 and thus we will know point $A = (\cos\phi, \sin\phi)$.

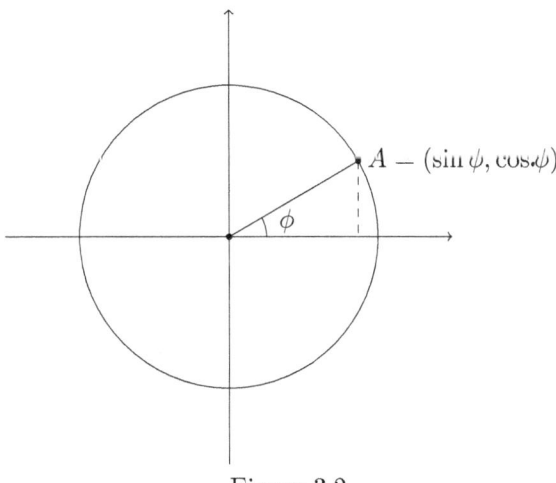

Figure 3.2

Taking into account the trigonometric formula of the triple angle $\sin 3\alpha = 3\sin\alpha - 4\sin^3\alpha$ (easily deduced by applying twice the sine formula of the sum angle - first with α and α to find $\sin 2\alpha$, then with α and 2α to find $\sin 3\alpha$) and that we are looking for the angle α such that $3\alpha = \phi$, we can write the equation:

$$4\sin^3\alpha - 3\sin\alpha + \sin\phi = 0$$

We want a method to trisect **any** angle, so let us suppose that we have an angle ϕ that meets the condition $\sin\phi = 3n/m$, where n and m are positive integers, with no common divisors, and both are not multiple of 3 (for example, angle ϕ such that $\sin\phi = 3/5$). So, for that angle, we are looking for a value of $\sin\alpha$ that:

$$4m\sin^3\alpha - 3m\sin\alpha + 3n = 0$$

It should be noted that knowing (and being able to construct with ruler-and-compass steps) $\sin\alpha$ and the angle α is completely equivalent. Therefore, we are trying to construct a value x which is solution of the equation:

(3.2) $$4mx^3 - 3mx + 3n = 0$$

If we can prove that the polynomial of equation (3.2) is irreducible, then we will come to the conclusion that the value x cannot be constructed with ruler-and-compass steps, since in the previous problem we saw that a constructible number with ruler-and-compass must be the solution of an irreducible power-of-two degree polynomial, and then it is not possible to be solution of an irreducible 3-degree polynomial that is also an irreducible power-of-two degree polynomial.

Therefore, if we show that polynomial $f(x) = 4mx^3 - 3mx + 3n$ is irreducible, then we can conclude that there is no method to trisect angles such that $\sin \phi = 3n/m$ in the indicated conditions and therefore that there is no method to trisect any angle.

But this polynomial is a perfect fit for Schonemann criterion, since it is primitive (m and n have no common divisors and m is not a multiple of 3), prime $p = 3$ divides all its coefficients except the highest degree one (again, because m is not a multiple of 3) and $p^2 = 9$ does not divide C_0 (because n is not multiple of 3).

The reasoning is completed: at least the angles such that $\sin \phi = 3n/m$, where n and m positive integers, with no common divisors and not multiples of 3, CANNOT be trisected with ruler-and-compass steps.

FINAL REMARKS

With this problem we have completed the explanation of unsolvability with constructions of ruler-and-compass of two of the three best known classical problems, although we have provided an alternative solution of both with the "trick" of a marked paper strip.

The third and famous related problem, also unsolvable with ruler-and-compass construction, was to try to "square the circle" (i.e., to construct with ruler-and-compass steps a segment of length π starting from a segment whose length is a rational number). The proof of its impossibility is more complicated, since we must prove that π is not a solution of any equation with rational coefficients (so, in particular, it is not a solution of an irreducible power-of-two degree polynomial). We will see that in the second part of this book, in the problem entitled "The transcendence of e and π".

The five regular solids

(Euclid – 290 BC)

PROBLEM

To prove that there are only five regular solids (the well-known tetrahedron, hexahedron, octahedron, dodecahedron and icosahedron).

HISTORY

The so-called regular solids were already known at the time of the Pythagoreans (around 600 BC), although we should still give credit of the Neolithic people who lived in what is now Scotland, who left us stone models that represented them and that can be seen at the "Ashmolean Museum" in Oxford.

Plato described them mathematically for the first time in his work "Timaeus" (around 350 BC), associating the tetrahedron with fire, the cube with the earth, the icosahedron with water, the octahedron with air and the dodecahedron with ether (what he thought was the substance that formed constellations and heavens). But it was Euclid who proved, in his immortal book "Elements", that there could not be any more.

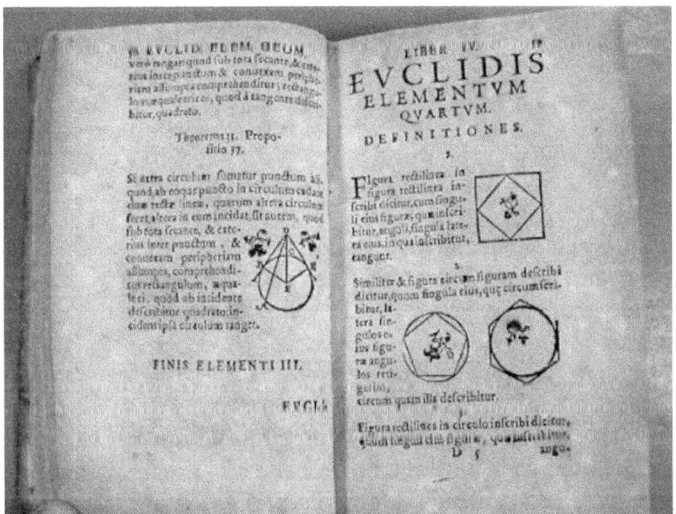

Fragment of the "Elements" – Euclid
Chapter IV – Latin translation

However, the following proof is much more modern than the one provided by Euclid, since part of it is based on the theorem of Euler, who lived in the eighteenth century.

SOLUTION

DEFINITION 4.1. *We define a **regular solid** as a convex polyhedron that satisfies the following properties:*

- *All faces are equal regular polygons.*
- *The number of faces converging at every vertex is the same.*

Example: The hexahedron (also known as cube) is a convex polyhedron (that is, if we take two of its points, the segment that joins them belongs entirely to it) whose faces are the same regular polygon (squares) and at whose vertices an identical number of faces converges (in this case, three). It is therefore a regular solid.

THEOREM 4.1. *(Euler) In a convex polyhedron, the number of faces F, the number of edges E and the number of vertices V meet the following equation:*

$$(4.1) \qquad\qquad F + V - E = 2$$

PROOF. Consider a convex polyhedron and remove one of its faces. What remains is a kind of "open box" where we can place something inside. Now, assuming that its edges can be lengthened or shortened with flexibility and that their faces are also formed by elastic material, we can "flatten" the box until the polyhedron lies in a plane. This is possible because the original polyhedron is convex by assumption.

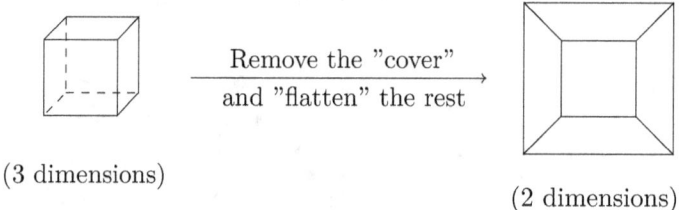

(3 dimensions)   Remove the "cover" and "flatten" the rest   (2 dimensions)

Figure 4.1

Let us try to calculate now the value of the formula $F + V - E$ in this figure, which we call a graph. First, suppose that in a face of the graph that is not a triangle we add a diagonal between two of its vertices. In this process we are creating a new edge as well as a new face (since the original face splits into two when divided by the new edge). Therefore, the value of $F + V - E$ of the original graph remains the same for the new graph (F has increased by one, while E has increased by one too, so we are adding one unit and subtracting another one in the value). We can now move on to the calculation of the value of $F + V - E$ for the new graph. Following the same reasoning, we repeat the process until all faces of the last graph are triangles, stopping there since there are no more diagonals to draw: the value of $F + V - E$ of the last graph (entirely formed by triangles) is the same as the original one.

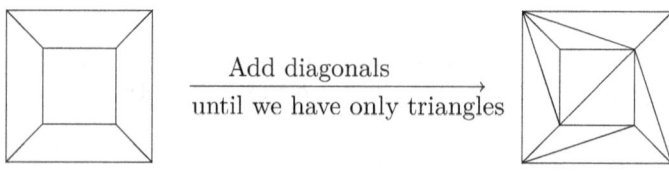

Add diagonals until we have only triangles

Figure 4.2

Now consider the following 2 operations:

a) In the graph, we remove a triangle that has an edge (and only one) that is shared with the rest of the triangles, creating another graph.

b) In the graph, we remove a triangle that has an edge (and only one) that is not shared with any other triangle, creating another graph.

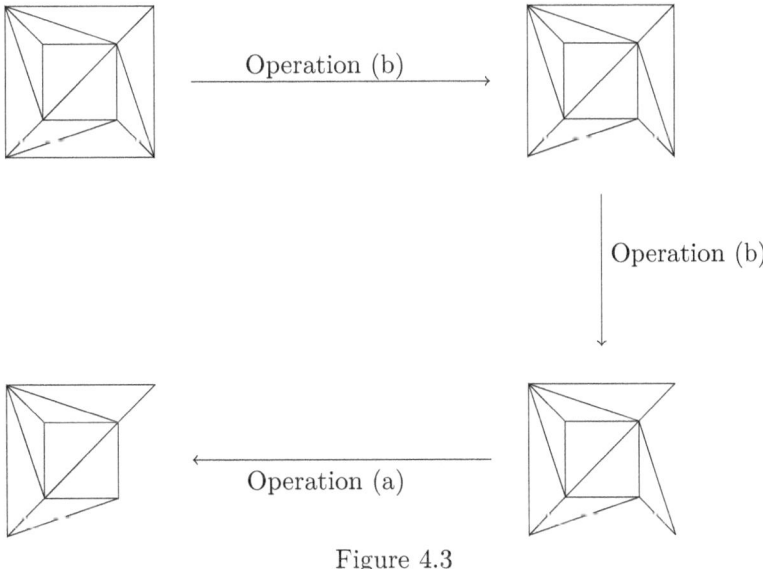

Figure 4.3

If we perform the operation (a) we are removing a vertex, a face and two edges of the graph, so the value of $F + V - E$ remains constant again. If we perform the operation (b) we remove an edge and a face: also here the value of $F + V - E$ is the same.

The idea is to change the graph with operations (a) and (b), taking care to perform operation (a) whenever possible, avoiding to previously perform operation (b) in that case. If we take this precaution, the process will always end the same way: with a graph where there is only one triangle (the proof of this fact is left to the reader). Since the value of $F + V - E$ has remained constant, it is only necessary to calculate it for the final graph (one triangle) to know its original value. In the case of the triangle, we have ($F = 1, V = 3, E = 3$), so $F + V - E = 1$.

However, recall that the first graph was the result of "flattening" the original polyhedron **after removing a face**, so you have to add a face in the original polyhedron calculations, finding the equation $F + V - E = 2$. $\square$

PROPOSITION 4.1. *In a regular solid, let e_v be the number of edges that converge at a vertex and let e_f be the number of edges on each face. Then:*

(4.2)
$$\frac{1}{e_f} + \frac{1}{e_v} - \frac{1}{2} = \frac{1}{E}$$

PROOF. Imagine a regular solid that has "exploded" from within, so that each face has been separated from the rest (see Figure 4.4) and each edge has turned into two edges (one edge for each face). If we imagine the regular solid in this way, we deduce that the number of faces by the number of edges of each face (in the example of the cube, 6 faces by 4 edges in each face = 24) has to be equal to twice the number of original edges (in the cube, 12 edges), due to the fact that we have commented before (the same edge is in each of the two faces that were together before the explosion, so each edge must now be counted twice). Then:

$$(4.3) \qquad\qquad F \cdot e_f = 2E$$

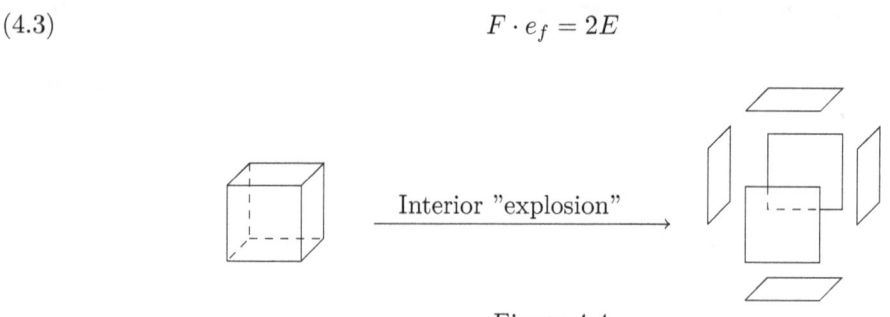

Figure 4.4

On the other hand, let us imagine now that all the edges of the regular solid have broken into two parts (as in Figure 4.5). In the new figure, each edge appears twice. Now, the number of vertices by the number of edges of each vertex (in the example of the cube, 8 vertices by 3 edges in each vertex = 24) is equal to twice the number of original edges (in the cube, 12 edges), since again we count each edge twice (one for each broken part). Therefore, it follows that:

$$(4.4) \qquad\qquad V \cdot e_v = 2E$$

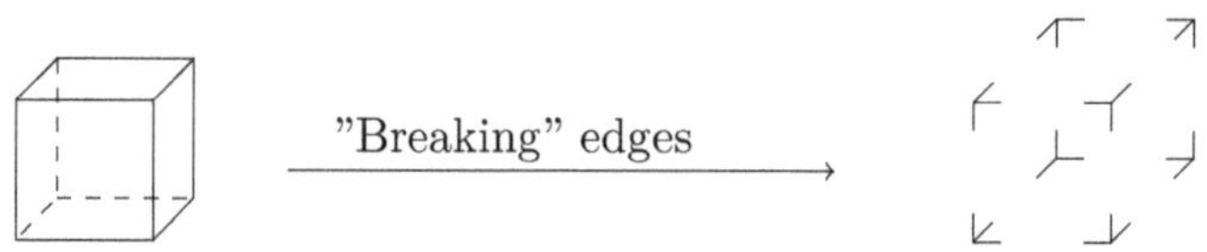

"Breaking" edges $\longrightarrow$

Figure 4.5

Substituting (4.3) and (4.4) in the formula of Euler's theorem (4.1):

$$F + V - E = 2 \quad \Rightarrow \quad \frac{2E}{e_f} + \frac{2E}{e_v} - E = 2 \quad \Rightarrow \quad \frac{1}{e_f} + \frac{1}{e_v} - \frac{1}{2} = \frac{1}{E}$$

$\square$

The previous formula is very restrictive, since it only allows very few combinations of E, e_f and e_v that meet the equality. In fact, there are only five possible combinations, as we will see below.

THEOREM 4.2. *(Euclid) There are only five regular solids.*

PROOF. Take the equality (4.2) for regular solids, where e_f and e_v are integers that are greater than 2 (it is clear that the number of edges of a face is an integer that is greater than 2 - a face must be, at least, a triangle -, and the number of edges in a vertex is an integer that is greater than 2 - there are no vertices where only 2 edges converge to form a polyhedron).

We have that $E > 0$, so we are looking for integers which are greater than 2 and meet the following inequality:

$$(4.5) \qquad\qquad \frac{1}{e_f} + \frac{1}{e_v} > \frac{1}{2}$$

There are only five integer possibilities which are greater than 2 and satisfy the inequality (2.5): values $(e_f, e_v) = \{(3,3), (4,3), (3,4), (5,3), (3.5)\}$. With greater values, fractions $1/e_f$ and $1/e_v$

decrease their value and they cannot add up to a number that is greater than $1/2$. The case $(3, 4)$, for example, indicates the possibility of the existence of a regular solid with $e_f = 3$ (three edges per face, that is, triangles) and $e_v = 4$ (four edges that reach the same vertex), that is, the octahedron.

e_f	e_v	E	F	V	Proposed name
3	3	6	4	4	Tetrahedron
4	3	12	6	8	Hexahedron
3	4	12	8	6	Octahedron
5	3	30	12	20	Dodecahedron
3	5	30	20	12	Icosahedron

The first two columns of the previous table are the different possibilities of e_f and e_v that satisfy the inequality (4.5); the third, fourth and fifth columns are the result of applying formulas (4.2), (4.3) and (4.4), respectively. Finally, the proposed name results from joining the Greek root equivalent to the number of faces ("tetra" is four, "hexa" is six, etc.) and the suffix "hedron" from Greek word "edron" = faces.

Therefore, as announced in the theorem, only these five regular solids can exist. □

FINAL REMARKS

In the previous proof we have shown that there can only be five regular solids, but it could happen that when trying to build any of them we would encounter a problem that we had not foreseen (maybe some other equation that relates the number of vertices, edges and faces that we have disregarded).

Actually, as our ancestors know, the five regular solids do exist. To verify this, we can give the coordinates of its vertices.

1. Tetrahedron

Put four vertices in the following coordinates:

$$V_1 = \left(0, 0, \frac{\sqrt{6}}{4}\right), V_2 = \left(\frac{\sqrt{3}}{3}, 0, -\frac{\sqrt{6}}{12}\right), V_3 = \left(-\frac{\sqrt{3}}{6}, \frac{1}{2}, -\frac{\sqrt{6}}{12}\right), V_4 = \left(-\frac{\sqrt{3}}{6}, -\frac{1}{2}, -\frac{\sqrt{12}}{12}\right)$$

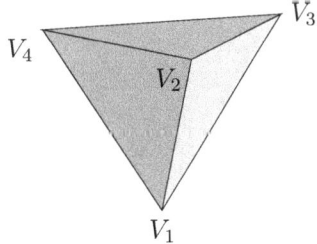

Figure 4.6

It can be verified that each vertex is at distance one unit from any other, and all of them are at distance $\sqrt{6}/4$ from the origin of coordinates. If we join all vertices with the rest, we have built the tetrahedron (4 faces, 4 vertices, 6 edges, 3 edges per face - triangles - and 3 edges per vertex).

2. Hexahedron

Put eight vertices in the following coordinates:

$$\left(\pm\frac{1}{2}, \pm\frac{1}{2}, \pm\frac{1}{2}\right)$$

Join each of the vertices with the other three that have common values in two coordinates, and build the hexahedron (6 faces, 8 vertices, 12 edges, 4 edges per face - squares - and 3 edges per vertex). Constructed edges have length one unit and all vertices are at distance $\sqrt{3}/2$ from the coordinate origin.

3. Octahedron

Put six vertices in the following coordinates:

$$V_1 = \left(0, 0, \frac{\sqrt{2}}{2}\right), V_2 = \left(\frac{1}{2}, \frac{1}{2}, 0\right), V_3 = \left(\frac{1}{2}, -\frac{1}{2}, 0,\right)$$

$$V_4 = \left(-\frac{1}{2}, \frac{1}{2}, 0\right), V_5 = \left(-\frac{1}{2}, -\frac{1}{2}, 0,\right), V_6 = \left(0, 0, -\frac{\sqrt{2}}{2}\right),$$

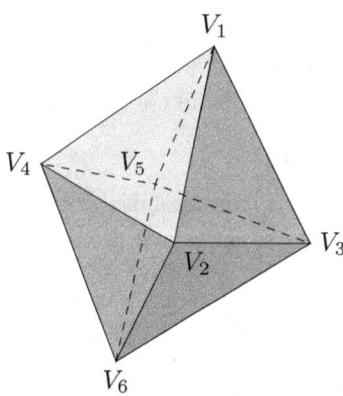

Figure 4.7

Let us join both V_1 and V_6 with V_2, V_3, V_4, V_5; and both V_2 and V_5 with V_3, V_4. This way we build the octahedron (8 faces, 6 vertices, 12 edges, 3 edges per face - triangles - and 4 edges per vertex). Constructed edges have length one unit and all vertices are at distance $\sqrt{2}/2$ from the coordinate origin.

4. Dodecahedron

Let ϕ be the golden ratio $\left(\phi = \frac{1+\sqrt{5}}{2}\right)$. Put 20 vertices in the following coordinates:

$$\frac{1}{2}\left(\pm\phi^2, \pm 1, 0\right), \frac{1}{2}\left(\pm 1, 0, \pm\phi^2\right), \frac{1}{2}\left(0, \pm\phi^2, \pm 1\right), \frac{1}{2}\left(\pm\phi, \pm\phi, \pm\phi\right)$$

Let us denote the 12 vertices with a coordinate equal to 0 by "type A" and the remaining 8 vertices by "type B". Let us now join each of the 12 vertices of "type A", for example $\frac{1}{2}\left(-1, 0, \phi^2\right)$, with the following three vertices:

- That "type A" vertex with which it shares the coordinate 0 in the same position but has opposite sign in the coordinate with value ± 1. In our example, vertex with coordinates $\frac{1}{2}\left(1, 0, \phi^2\right)$
- Those two "type B" vertices whose sign in the positions where it does not have the coordinate equal to 0 is the same. In our example, these vertices are $\frac{1}{2}\left(-\phi, \phi, \phi\right)$ and $\frac{1}{2}\left(-\phi, -\phi, \phi\right)$

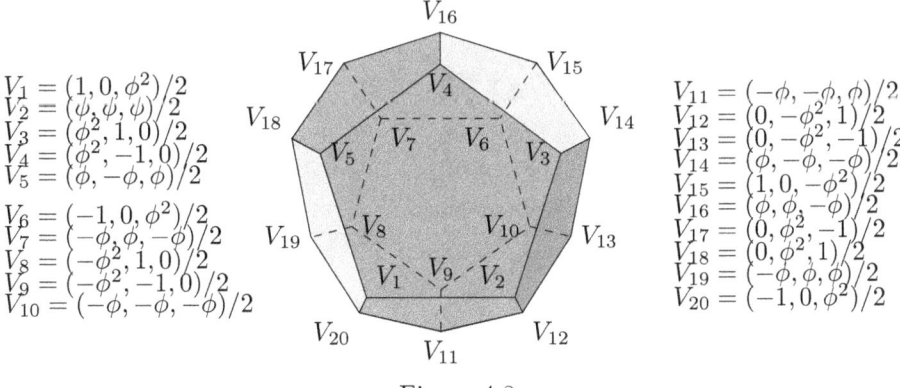

$V_1 = (1, 0, \phi^2)/2$
$V_2 = (\phi, \phi, \phi)/2$
$V_3 = (\phi^2, 1, 0)/2$
$V_4 = (\phi^2, -1, 0)/2$
$V_5 = (\phi, -\phi, \phi)/2$

$V_6 = (-1, 0, \phi^2)/2$
$V_7 = (-\phi, \phi, -\phi)/2$
$V_8 = (-\phi^2, 1, 0)/2$
$V_9 = (-\phi^2, -1, 0)/2$
$V_{10} = (-\phi, -\phi, -\phi)/2$

$V_{11} = (-\phi, -\phi, \phi)/2$
$V_{12} = (0, -\phi^2, 1)/2$
$V_{13} = (0, -\phi^2, -1)/2$
$V_{14} = (\phi, -\phi, -\phi)/2$
$V_{15} = (1, 0, -\phi^2)/2$
$V_{16} = (\phi, \phi, -\phi)/2$
$V_{17} = (0, \phi^2, -1)/2$
$V_{18} = (0, \phi^2, 1)/2$
$V_{19} = (-\phi, \phi, \phi)/2$
$V_{20} = (-1, 0, \phi^2)/2$

Figure 4.8

This way we build the dodecahedron (12 faces, 20 vertices, 30 edges, 5 edges per face - pentagons - and 3 edges per vertex). Constructed edges have length one unit and all vertices are at distance $\sqrt{6} \cdot \left(\sqrt{3 + \sqrt{5}}\right)/4$ from the coordinates origin.

5. Icosahedron

Put 12 vertices in the following coordinates:

$$\frac{1}{2}\left(\pm\phi, 0, \pm 1\right), \frac{1}{2}\left(\pm 1, \pm\phi, 0\right), \frac{1}{2}\left(0, \pm 1, \pm\phi\right)$$

Let us now join each of the 12 vertices, for example $\frac{1}{2}\left(-1, \phi, 0\right)$, with the following 5 vertices:

- That vertex with which it shares the coordinate 0 in the same position but has opposite sign in the coordinate with value ± 1. In our example, vertex with coordinates $\frac{1}{2}\left(1, \phi, 0\right)$.
- Those two vertices in which the unit (with its sign) becomes ϕ (with the same sign) and in which ϕ (with its sign) becomes 0. In our example, vertices with coordinates $\frac{1}{2}\left(-\phi, 0, \pm 1\right)$.
- Those two vertices in which ϕ (with its sign) becomes the unit (with the same sign) and in which the unit (with its sign) becomes 0. In our example, vertex with coordinates $\frac{1}{2}\left(0, 1, \pm\phi\right)$.

This way we build the icosahedron (20 faces, 12 vertices, 30 edges, 3 edges per face - triangles - and 5 edges per vertex). Constructed edges have length 1 and all vertices are at distance $\sqrt{10 + 2\sqrt{5}}/4$ from the coordinates origin.

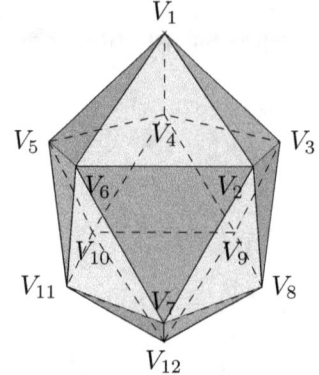

$V_1 = (\phi, 0, 1)/2$
$V_2 = (1, \phi, 0)/2$
$V_3 = (0, 1, \phi)/2$
$V_4 = (0, -1, \phi)/2$
$V_5 = (1, -\phi, 0)/2$
$V_6 = (-1, 0, \phi^2)/2$

$V_7 = (0, 1, -\phi)/2$
$V_8 = (-1, \phi, 0)/2$
$V_9 = (-\phi, 0, 1)/2$
$V_{10} = (-1, -\phi, 0)/2$
$V_{11} = (0, -1, -\phi)/2$
$V_{12} = (-\phi, 0, -1)/2$

Figure 4.9

Chapter 5

Distance and radius of the Moon and the Sun

(Aristarchus – 260 BC)

PROBLEM

Taking the radius of Earth as a reference, to calculate the distances Earth - Moon and Earth - Sun, as well as the radii of the Moon and the Sun.

HISTORY

Aristarchus of Samos (310 BC - about 230 BC), was the first scientist to propose the heliocentric model of the Universe, although his ideas were rejected for centuries in favor of the geocentric theory of Aristotle and Ptolemy.

Only 1800 years later, thanks to the formulas of Kepler and Newton, his model was finally recognized and he came to occupy a prominent place in the history of Astronomy. However, it is logical to assume that during that time many scientists believed in that model, although they did not publicly defend it out of fear of being accused of heresy.

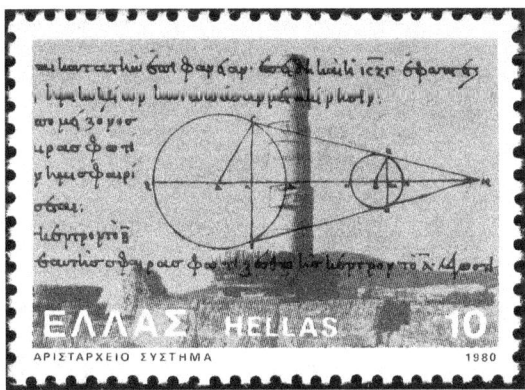

Commemorative Greece stamp of Aristarchus

Aristarchus devised the following reasoning to calculate the values of the lunar and solar radii, as well as their distances to Earth (taking the unknown terrestrial radius as a reference).

SOLUTION

First observation: "Equal apparent sizes of the Sun and the Moon"

One of the first measurements made by Aristarchus is the angle that the Sun and the Moon occupies in the sky for an observer on Earth. Interestingly, this value is the same for both, since the apparent size seen from our planet is identical; that phenomenon can be seen at any time (measuring the size of both in the sky), but it is spectacular to watch during a total solar eclipse, when the Moon hides the Sun until they overlap completely (causing a nice visual effect for a few seconds, when only the solar contour and the continuous eruptions on its surface can be seen from Earth).

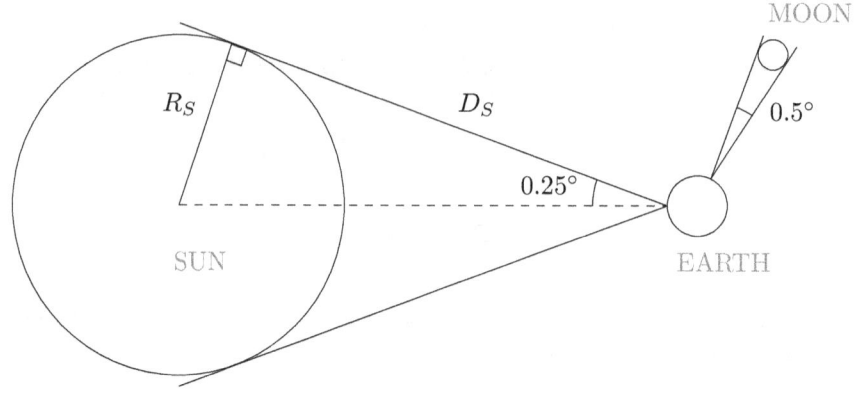

Figure 5.1

Aristarchus set as 0.5 degrees the angle from which an Earth observer contemplates both the Sun and the Moon (see figure 5.1). Let R_S and R_M be the radii of the Sun and the Moon, and D_S and D_M be the distances of the Sun and the Moon from Earth. Then, the measurement of Aristarchus can be written as:

$$\frac{R_S}{D_S} = \frac{R_M}{D_M} = \tan\left(0.25 \cdot \frac{\pi}{180}\right)$$

Since the angle (in radians) $(0.25 \cdot \pi/180$ is a value that is very close to 0, its tangent can be approximated by its value $(\tan(x) \approx x$ if x is close to 0), and we have that we can write the equation as:

(5.1)
$$\frac{R_S}{D_S} = \frac{R_M}{D_M} = 0.00436$$

In figure 5.1 (which obviously is not to scale) it might seem that we have cheated by placing the observer on Earth at the point that suits us the most (closest point to the Sun and the Moon, respectively). In fact, the enormous distances to the Sun and the Moon (compared to the radius of Earth) make no difference in the calculations. However, to be strict, we should make observations at the moment in which the Sun (respectively, the Moon) is at the midpoint of its daily trajectory: for the Sun, at noon, and for the Moon, we make the observations during all the time that the Moon has been in the sky in one day and we take the observation that occurred in mid time.

Insisting on the same subject: the fact that the proportion between radii and distance to Earth is the same for the Moon and for the Sun was such a coincidence that it likely impressed the defenders of geocentricism (Earth as the center of the sun's and other known planets orbits) so much that they ignored many other evidences that suggested otherwise.

Second observation: "Quotient between distances to the Moon and to the Sun"

Aristarchus knew that the different phases of the Moon could be explained by the relative position between the Sun, the Moon and Earth. What we call "full moon" occurs when the Moon is in the opposite direction to the Sun (relative to Earth), hence receiving its visible face (from Earth) all solar light; most of the time (except during eclipses) Earth does not cover that light.

30

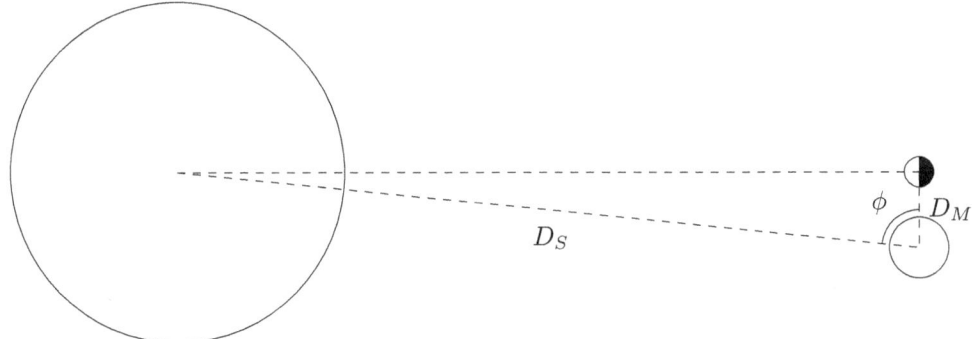

Figure 5.2

What we call "new moon" occurs when the Moon and the Sun are in the same direction and we see the visible face of the Moon (from Earth) totally dark.

However, Aristarchus was interested in the moment in which there was "half moon", that is, when you can see from Earth exactly half of its visible face illuminated by the Sun and the other half dark. Aristarchus supposed that this happened when the Sun - Moon - Earth angle was 90°, as shown in figure 5.2. When that position occurred, he tried to experimentally calculate angle ϕ (Moon - Earth - Sun angle) in a rudimentary way, observing where the Sun and the Moon were in the sky at that moment. That angle ϕ is very close to 90° (because the Sun is much farther away from Earth than the Moon) and he measured it as 87°, from which he deduced that:

$$\cos(87°) = \frac{D_M}{D_S}$$

Aristarchus did not have precise tables of cosines, but he approximated the value of $\cos(87°)$ with the value of the inverse of 19.1, which means that he found that Sun - Earth distance was 19.1 times greater than Moon - Earth distance.

(5.2) $$D_S = 19.1 \cdot D_M$$

Applying what we have found in (5.1), we also deduce that:

(5.3) $$R_S = 19.1 \cdot R_M$$

Third observation: "Total Eclipse of the Moon"

Finally, Aristarchus had a great idea that allowed him to calculate another quotient between distances that interested him. He used the time it took for the Moon to cross the shade of Earth during a total lunar eclipse.

During a total lunar eclipse (figure 5.3), suppose that the distance the Moon has to travel in the shade generated by Earth is $2d$ (the trajectory followed by the Moon is circular, of course, but at that time there was no better approximation than to assume it to be linear). Aristarchus realized that the Moon took approximately 160 minutes to travel that distance (from the moment the Moon begins to disappear in the shade until the moment that begins to reappear) and anyone contemplating a total lunar eclipse in the future will be able to corroborate this result.

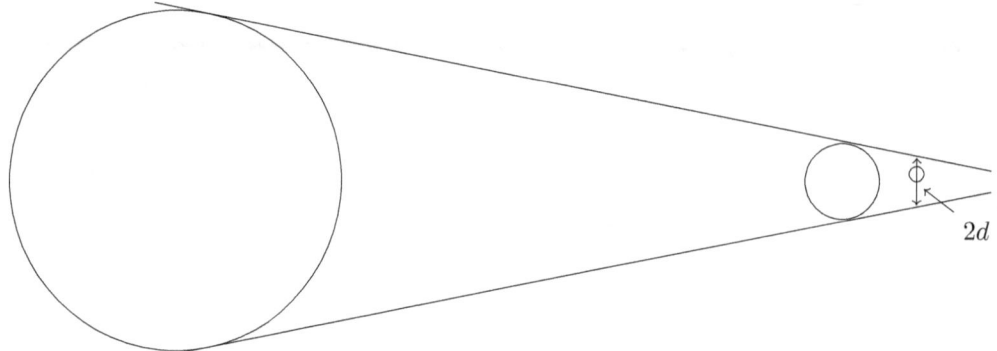

Figure 5.3

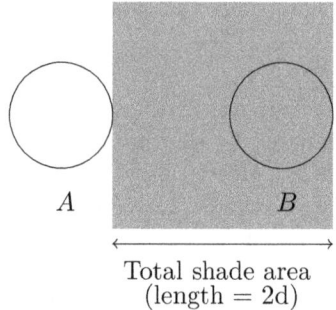
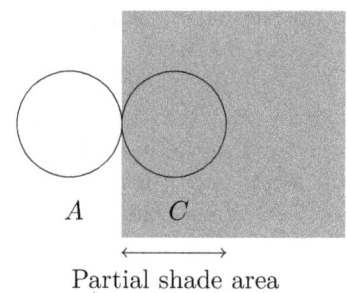

Total shade area
(length = 2d)

Partial shade area
(length = $2R_M$)

Time $A \to B = 160$ minutes   Time $A \to C = 60$ minutes

Figure 5.4

Moreover, Aristarchus calculated the time that the Moon takes to travel a distance equal to its diameter (from the moment the Moon begins to disappear in the shade until it begins to disappear completely), resulting in a value of 60 minutes, approximately. Then, by simple proportion:

$$(5.4) \qquad \frac{2d}{2R_M} = \frac{160}{60} \qquad \Rightarrow \qquad d = \frac{8}{3} \cdot R_M$$

Final calculations

With the previous results, Aristarchus established the final model to calculate the distance Earth - Moon and Earth - Sun depending on the radius of Earth (R_E). He used a model based in figure 5.5.

By similarity of triangles, it follows that:

$$\frac{D}{R_E} = \frac{D - D_M}{d} \qquad \text{and} \qquad \frac{D}{R_E} = \frac{D + D_S}{R_S}$$

Both equations can be written (isolating D in each of them) as:

$$\frac{D}{R_E} = \frac{D_M}{R_E - d} \qquad \text{and} \qquad \frac{D}{R_E} = \frac{D_S}{R_S - R_E}$$

Equalizing both equations:

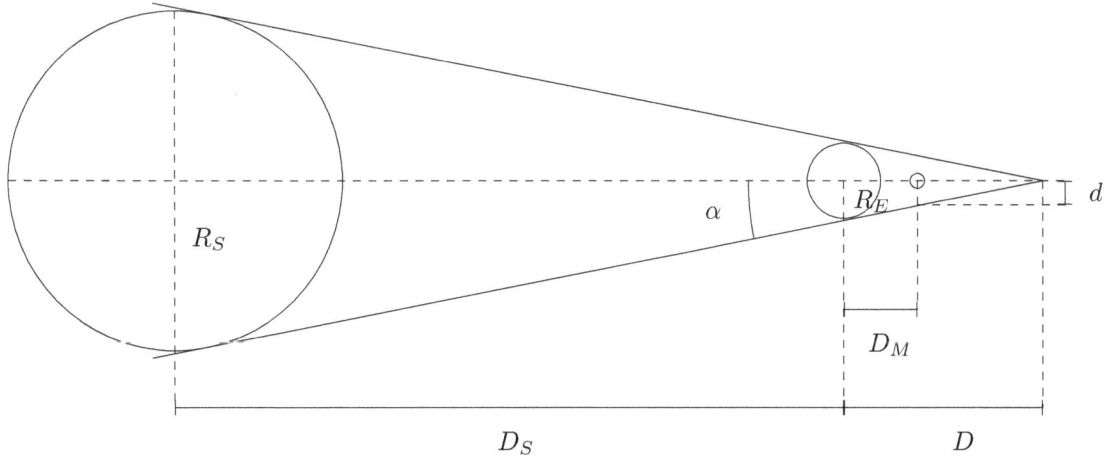

Figure 5.5

(5.5)
$$\frac{D_M}{R_E - d} = \frac{D_S}{R_S - R_E}$$

If we replace in this equation what we found in equation (5.2) and (5.4):

$$\frac{1}{R_E - (8/3) \cdot R_M} = \frac{19.1}{R_S - R_E}$$

and the relation found in (5.3):

$$\frac{1}{R_E - (8/3) \cdot (R_S/19.1)} = \frac{19.1}{R_S - R_E}$$

so we find that $R_S/R_E = 5.48$, i.e., the radius of the Sun is about 5.48 times greater than the radius of Earth. And, therefore, $R_E/R_S = 1/5.48$, leading to, by applying (5.3), to $R_E/R_M = (1/5.48) \cdot 19.1 = 3.48$, that is, the radius of Earth is about 3.48 times greater than the radius of the Moon.

To calculate the distances to the Sun and to the Moon, we use (5.1) in the results $R_S/R_E = 5.48$ and $R_E/R_M = 3.48$, to infer that $0.00436 D_S/R_E = 5.48$ and $R_E/(0.00436 D_M) = 3.48$. Operating, the conclusion is $D_S/R_E = 1256.88$ (the distance Earth - Sun is 1256.88 times greater than the radius of Earth) and $D_M/R_E = 65.90$ (the distance Earth - Moon is 65.90 times greater than the radius of Earth).

Let us see in a table the real values and those calculated by Aristarchus:

Measure	Real value	Aristarchus
(Sun radius) / (Earth radius)	109	5.48
(Earth radius) / (Moon radius)	3.50	3.48
(Distance Earth − Sun) / (Earth radius)	23500	1256.88
(Distance Earth − Moon) / (Earth radius)	60.32	65.90

FINAL REMARKS

As we can see, the calculations relative to the Moon (its radius and its distance to Earth) were calculated with amazing precision, while those relative to the Sun had a great error, especially due to the difficulty of calculating the angle ϕ of our second observation.

Errors made by Aristarchus:

- Angle of the first observation: the actual value is $0.53°$ instead of the value of $0.50°$ calculated by him.
- Angle of the second observation: the actual value is $89.5°$ instead of the value of $87°$ calculated by him.
- Probably Aristarchus did not have a good approximation of the value of π, so the value found in the formula (5.1) was not so precise.
- He approximated the circular trajectory of the Moon (when it passes through the shade of Earth during an eclipse) by a linear path.

All the values that Aristarchus calculated were in relation to the terrestrial radio, which was not known with precision at that time. As we will see in a later problem, Eratosthenes achieved a good approximation a few years later.

Chapter 6

Approximate the value of π

(Archimedes – 250 BC)

PROBLEM

To find a geometric method to calculate the number π to a high level of accuracy.

HISTORY

Archimedes was a mathematician, physicist, engineer, inventor and astronomer that was born in the Sicilian city of Syracuse (about 287 BC). He died in that same place in 212 BC, when the Roman forces invaded it during the Second Punic War. Apparently, a Roman soldier killed him while he was trying to solve a mathematical problem, despite the fact that General Marco Claudio Marcelo, who was aware of his well-deserved fame, had commanded not to harm him.

"Death of Archimedes"
Engraving of Gustave Courtois (1853 – 1923)

Although his mathematical achievements were very numerous, measuring the perimeter of the circle or, equivalently, approximating the number π, is his best-known contribution.

SOLUTION

Let us take a look at figure 6.1: we have a circle of diameter d with an inscribed hexagon in it and a circumscribed hexagon around it. If we calculate the perimeters of both hexagons, it is clear that the length of the circumference will be a value that is between both.

Let us begin by calculating (see figure 6.2) the lengths of the regular inscribed and circumscribed hexagons; with this calculation, we will obtain our first approximation to the value of π.

The case of the inscribed hexagon (first graph of figure 6.2) is very simple. Since it has six sides, we can divide the hexagon into six triangles, so that in each of them the inner angle α has a value

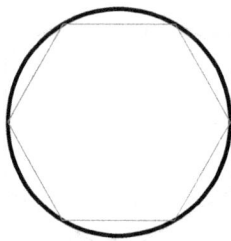

Figure 6.1

equal to 60°; moreover, these triangles have two sides with a length that equals the radius, which means that the triangle is equilateral. From this we infer that the length of a side of the hexagon (l_i) is $l_i = d/2$ and the perimeter of the hexagon (p_i) is $p_i = 6 \cdot l_i = 3d$.

The case of the circumscribed hexagon (second graph of figure 6.2) is not much more complicated. Here we can divide the hexagon (with length of a side equal to l_c and perimeter equal to p_c) into triangles and the inner angle α has a value that is equal to 60° again; if we take a half of any triangle, it has a side whose value equals the radius of the circle and has an angle of value equal to 30°, so $\tan(30°) = l_c/d$. Hence we infer that $(\sqrt{3})/3 = l_c/d$, that is, $l_c = d\sqrt{3}/3$ and the perimeter of the hexagon is $p_c = 6 \cdot l_c = 2\sqrt{3}d$.

The length of the circumference (equal to πd) is a value between these two numbers; so we get the first approximation of π: $3 < \pi < 2\sqrt{3} = 3.464....$

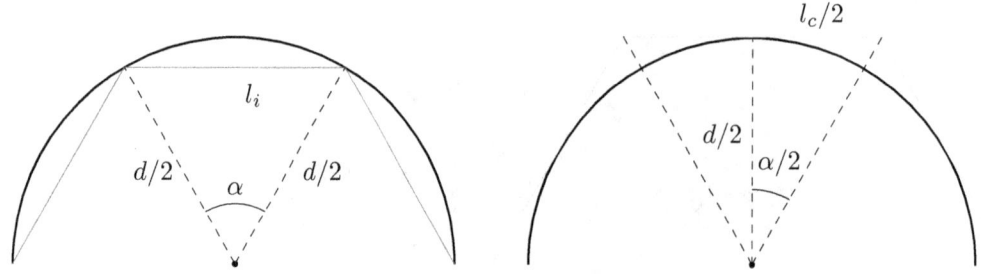

Figure 6.2

If now we use a a polygon with a greater number of sides (for example, a polygon with 12 sides) instead of an hexagon, the procedure is similar and it is still true that the length of the circumference is a value that is greater than the perimeter of the inscribed polygon but smaller than the circumscribed one. The problem is that the calculations of the perimeters are not as simple as those involving the hexagon.

Then, the objective is to calculate the perimeters of the regular circumscribed and inscribed polygons with a large number of sides in a simple manner. If we obtain that, we will find the value of the length of the circumference with an error ϵ that will be as small as we want. Archimedes' achievement was to find a method by which the perimeters of those regular n-gons of many sides can be easily calculated. This method, called Archimedes' algorithm, is based on two recursive formulas that we will now infer.

In figure 6.3, let O be the center of the circle and $\overline{AB} = 2t$, $\overline{CD} = 2s$ be the length of the sides of the circumscribed and inscribed polygons of n sides, respectively. Let M be the midpoint of AB, N be the midpoint of CD and Z be the point of intersection of the tangent to the circle at point C with segment MA. It follows that $\overline{ZM} = \overline{ZC}$ ($\overline{ZM}$ is the tangent to the circumference at point M and $\overline{ZC}$ is the tangent at point C, so by symmetry Z cannot be closer to one of the points than to the other). Set t' to be this distance, which is equal to half of the length of the side

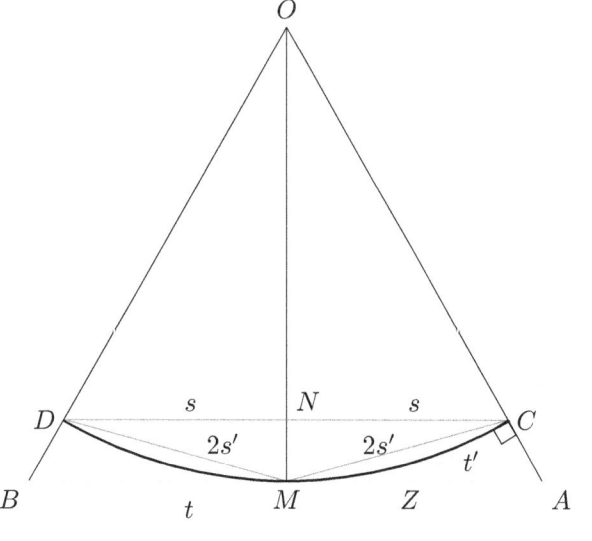

Figure 6.3

of the circumscribed regular polygon of $2n$ sides. It is also true that $\overline{MC} = \overline{MD}$ and denote this value as $2s'$, which corresponds to the length of the side side of the regular inscribed polygon of $2n$ sides.

As triangles ACZ and AMO are similar (they share the same angle at the point A and they both have a right angle - at C and at M, respectively -), then:

$$(6.1) \qquad \frac{t'}{t - t'} = \frac{\overline{ZC}}{\overline{ZA}} = \frac{\overline{MO}}{\overline{AO}}$$

It is also inferred, now by similarity of the triangles ODC and OBA, that:

$$(6.2) \qquad \frac{s}{t} = \frac{\overline{NC}}{\overline{MA}} = \frac{\overline{CO}}{\overline{AO}}$$

Since the right-hand side of equations (6.1) and (6.2) is identical (because $MO = CO$), then:

$$(6.3) \qquad \frac{t'}{t - t'} = \frac{o}{t} \qquad \Rightarrow \qquad t' = \frac{t \cdot o}{t + s}$$

Moreover, CMD and CZM are similar triangles (the angle at C of the triangle CNM - and, therefore, of triangle CMD - has the same value as the angle at M of the triangle CZM, and both triangles are isosceles), so:

$$(6.4) \qquad \frac{2s'}{2s} = \frac{t'}{2s'} \qquad \Rightarrow \qquad 2(s')^2 = s \cdot t'$$

Now, set C_n to be the perimeter of the circumscribed regular polygon of n sides and set I_n to be the perimeter of the inscribed one. Then:

$$C_n = 2nt \qquad I_n = 2ns \qquad\qquad C_{2n} = 4nt' \qquad I_{2n} = 4ns'$$

If we use the above equations together with those ones found in (6.3) and (6.4) we can eliminate the values of t, t', s, s' and find the *Archimedes' recurrence formulas*:

$$C_{2n} = \frac{2C_n \cdot I_n}{C_n + I_n} \qquad\qquad I_{2n} = \sqrt{I_n \cdot C_{2n}}$$

If we begin with two known values, C_6 and I_6 (values for the hexagon), we can apply the recurrence formulas as many times as we want, finding the values of C_{12}, I_{12}, C_{24}, I_{24} and so on, and we can calculate π to a level of accuracy that is as high as we want.

So, take the values that were already calculated for C_6 and I_6:

$$C_6 = 2\sqrt{3}d \qquad\qquad I_6 = 3d$$

where d is the diameter of the circle. Then we can calculate the values of C_{12}, I_{12}, ..., etc. If we stop at C_{96} and I_{96}, we find the values $C_{96} = d \cdot 3,142715...$ and $I_{96} = d \cdot 3,141032...$, where we find the approximation

$$3,141032 < \pi < 3,142716$$

FINAL REMARKS

Archimedes did not actually use square roots in his calculations, but approximations of them, so the precision he got was not as good as that shown in the solution. He was satisfied with the approximation $3 + 10/71 < \pi < 3 + 1/7$. However, the truth is that this was the first mathematical method used to approximate π that did not merely consist in performing an inaccurate measurement, as it had been the case up to that point.

Chapter 7

Calculation of the radius of Earth

(Eratosthenes – 245 BC)

PROBLEM

To calculate the radius of Earth.

HISTORY

Although it was believed at first that Earth was flat and the sky was a huge vault that enclosed it, some mathematicians in ancient Greece such as Pythagoras or Aristotle began to spread the thought of Earth was spherical. Pythagoras believed so because he considered the sphere to be the geometrically perfect body and a creation of God should meet such perfection. Aristotle relied both on the observations of the partial eclipses of the Moon (it can be seen that Earth overlaps it and that the edge of the shade is circular) and on the fact that the polar star is closer to the horizon as one travels to the South (which invalidated the theory of a flat Earth).

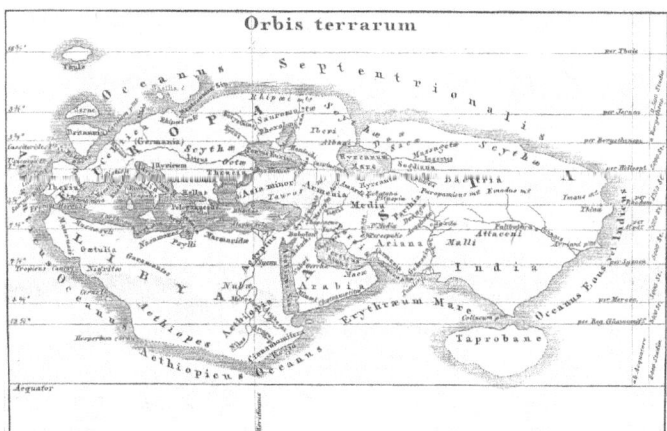

Map of the known world (Eratosthenes)
Reconstruction made in the 19th century

However, the first attempts of the Greeks to compute the radius of Earth gave rise to approximations with a high error. It was not until the middle of the third century BC when the Greek Eratosthenes, who lived in Egypt, devised a new method with which he managed to approximate the real value to a high level of accuracy. Eratosthenes is also known for the algorithm in order to determine prime numbers, which bears his name, and for the map that he drew of the world known at the time.

SOLUTION

Eratosthenes knew that on the summer solstice (the longest day in the Northern Hemisphere) at noon the sun's rays managed to reach the bottom of a deep well that existed near the actual city of Aswan, famous for its proximity to the Abu Simbel temples. That means in mathematical terms that the sun's rays were perpendicular to the ground in Aswan at that precise moment.

However, on the same day at the same time in the city of Alexandria, this phenomenon did not occur. In fact, if a stick was poked vertically into the ground, the shadow that it projected, although minimal, was not canceled (in Aswan, the stick would not have produced any shadow at all). This means that the angle α between the rays of the Sun and the vertical in Alexandria was not equal to 0.

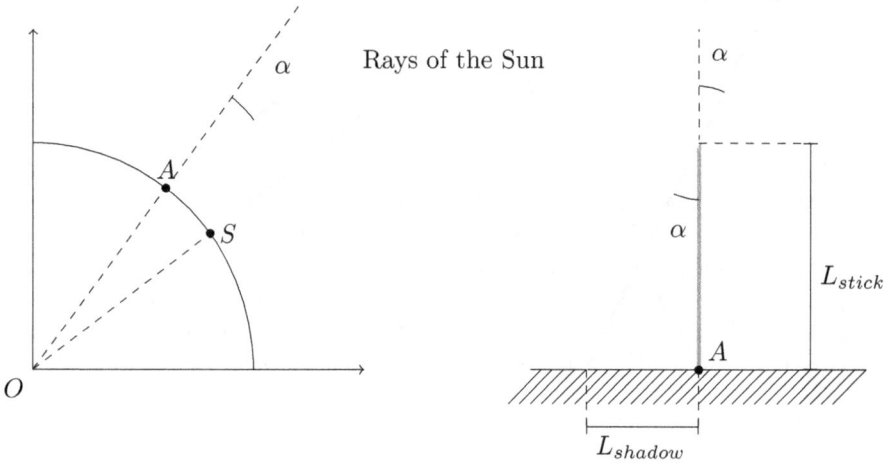

Figure 7.1

Let us have a look at the first graph of figure 7.1, where a part of a terrestrial meridian has been drawn: let O be the center of Earth, A be the city of Alexandria and S be the city of Aswan (Alexandria and Aswan are not in the same meridian, but the deviation is negligible when compared to the calculation of angle α, for example). The angle α, as already defined, is the angle between the vertical of Alexandria (the extension of the radius $\overline{OA}$) and the sun's rays (parallel at that time to the vertical of Aswan, or the extension of the radius $\overline{OS}$).

In the second graph of figure 7.1 a vertical stick poked into the ground helps us in the calculation of the angle α. If we calculate the length of the stick and the length of the shadow, we can infer that:

$$\tan \alpha = \frac{L_{shadow}}{L_{stick}}$$

The experimental calculation of α made by Eratosthenes led him to a value equal to 1/50-th of the circumference, or about $7°12'$.

Moreover, the legend says that Eratosthenes sent an expedition from Alexandria to Aswan to calculate the exact distance (in a straight line) between both cities, and they described the value as "5000 stadiums". If we consider that a "stadium" is equivalent to the Greek stadium of 185 meters, that means a distance of about 925 km.

Now we just need to apply a simple proportion: if the angle $\widehat{AOS}$ has a value equal to 1/50-th of the terrestrial meridian and that leads us to an arc length AS (calculated on the meridian) of 925

Km, then the total terrestrial meridian is 50 times this value, that is, 46250 km. In fact, Earth is not a perfect sphere and the length of the meridian is approximately 40000 km, so the error of the calculation of Eratosthenes was only 15%, which is surprising considering the inaccuracy of the measurements of that time.

The calculated terrestrial radius is the division of the longitude of the meridian by 2π, which would give us a value of about 7360 km (the real value is about 6370 km).

FINAL REMARKS

The errors in the measurements of Eratosthenes were:

- Aswan is not exactly in the Tropic of Cancer (latitude $23°43'$ at that time), where the Sun actually falls in the vertical on the summer solstice at noon, but a little further North (latitude $24°05'$).
- The α angle between the vertical of Alexandria and the solar rays on the summer solstice at noon at that time was approximately equal to 1/48-th part of the circumference, and not equal to 1/50-th as it was calculated.
- The actual distance between Alexandria and Aswan is 842 km and there is a difference of about $3°$ between the meridians that pass through each of these cities.

Chapter 8

Area of the section of a parabola

(Archimedes – 240 BC)

PROBLEM

To calculate the area of the section of a parabola.

HISTORY

There are many legends about Archimedes, the best mathematician of antiquity. A very well-known one assures that during the siege of Syracuse that caused his death he repelled an attack of the Roman fleet by focusing enormous mirrors towards their ships and provoking devastating fires with them using the reflection of the rays of the Sun. Unfortunately, this legend is unlikely to be true, since some students from the prestigious Massachusetts Institute of Technology repeated the experiment in the 21st century for a television network, and they prove that the attack could have only been successful on a very sunny day, early in the morning (due to the position of the Sun with respect to the port of Syracuse) and only if the ships remained motionless for ten minutes, all of which seems quite unlikely.

Archimedes using mirrors against Roman fleet
Painting of Giulio Parigi (1599)

Fortunately, the works that Archimedes left for posterity are no legend. The "quadrature of the parabola" is among them, where he described a method to calculate the area of a region delimited by it. The study of the parabola is one of the most important achievements of Archimedes: it was finished around 240 BC and it is based on the properties of the so-called "Archimedean triangles".

SOLUTION

Let us suppose that we have a parabola section, defined by two of its points A and B. Let D be the axis of the parabola, F be its focus and D_A (resp. D_B) be the perpendicular projection of A (resp. B) onto line D, as seen in figure 8.1.

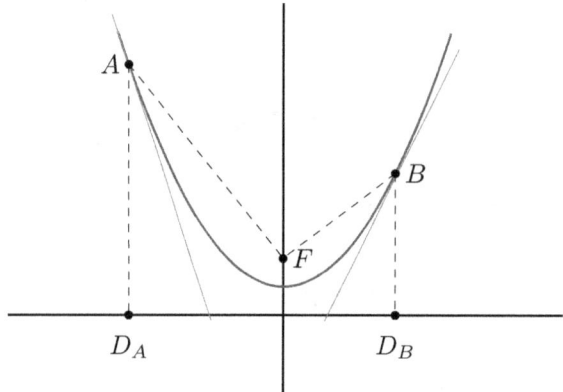

Figure 8.1

LEMMA 8.1. *The bisector of the angle $\widehat{FAD_A}$ (resp., $\widehat{FBD_B}$) is the tangent to the parabola at point A (resp., point B).*

PROOF. Point A satisfies that its distance to the directrix axis (line D) is the same as the distance to the focus of the parabola (point F), by definition of the parabola. Therefore, the triangle AFD_A is isosceles and the bisector that passes through A is perpendicular to segment $\overline{FD_A}$, which means that it is also the mediatrix of segment $\overline{FD_A}$ and therefore all of its points are at equal distance to points F and D_A.

Now, any point of this bisector that is **distinct** from A satisfies that it is at the same distance from F and D_A; however, is NOT at the same distance from F and from the directrix axis, since its distance to D is different than its distance to D_A (it only occurs that A has its projection to D in the point D_A). So, any point of this bisector other than A does not belong to the parabola (it does not fulfill its essential property). If we have a line that passes through a point of the parabola (point A) but does not cut any other point of it, then it is necessarily the tangent of the parabola at that point. We can apply the same reasoning for point B. □

DEFINITION 8.1. *Let A and B be any two points of a parabola, and let S be the intersection of the two tangents to the parabola at A and B. We say that the triangle that has points A, B and S as vertices is an **Archimedean triangle** and we say that the side $\overline{AB}$ is its base.*

LEMMA 8.2. *The median to the base of an Archimedean triangle is parallel to the axis of the parabola.*

PROOF. Let us see figure 8.2. Since S belongs to the tangent at A, we know by lemma 8.1 that it is a point of the bisector of the triangle AFD_A at vertex A, which means that the distances $\overline{SF}$ and $\overline{SD_A}$ are identical. By the same reasoning, the distances $\overline{SF}$ and $\overline{SD_B}$ are also equal, from which it follows that $\overline{SD_A} = \overline{SD_B}$, that is, S belongs to the mediatrix of the segment $\overline{D_AD_B}$. That means that the line that passes through S and is perpendicular to line D is the mediatrix of the $\overline{D_AD_B}$ segment. This mediatrix cuts the $\overline{AB}$ segment at its midpoint M. In other words, the segment $\overline{SM}$, which is parallel to the axis of the parabola, is the median to the side $\overline{AB}$ of the triangle. □

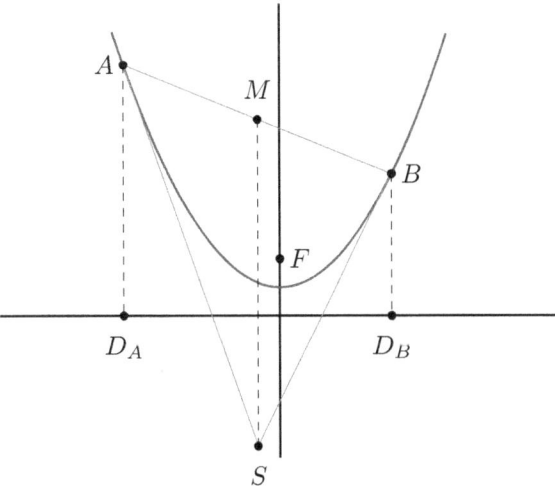

Figure 8.2

LEMMA 8.3. *With the notations used so far, the midpoint of the segment $\overline{SM}$ is a point of the parabola. Furthermore, the tangent to the parabola at this point is parallel to the segment $\overline{AB}$ and cuts the segments $\overline{AS}$ and $\overline{BS}$ at their midpoints.*

PROOF. Let O be the intersection of the segment $\overline{SM}$ and the parabola (we want to prove that O is the midpoint of $\overline{SM}$). Let us suppose that we draw the tangent to the parabola at O and let A' (resp. B') be the intersection of this tangent and the segment $\overline{SA}$ (resp. $\overline{SB}$); we want to prove that $\overline{AA'} = \overline{A'S}$ (resp., $\overline{BB'} = \overline{B'S}$). By definition, both $AA'O$ and $BB'O$ are Archimedean triangles (both are formed by two tangents to the parabola at two points and the segment that joins both points). By lemma 8.2, the medians to their bases are parallel to the axis of the parabola and therefore they are parallel to segment $\overline{SO}$; that is, the line that passes through A' (resp., B') and it is parallel to $\overline{SO}$ cuts the segment $\overline{AO}$ (resp., $\overline{BO}$) at its midpoint.

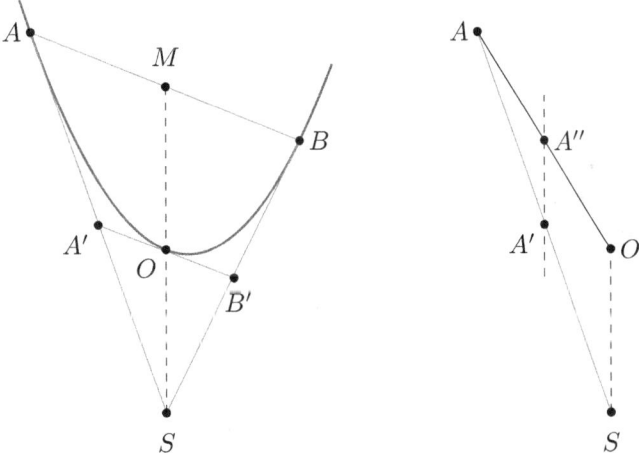

Figure 8.3

In figure 8.3, let A'' be the midpoint of segment $\overline{AO}$ so that the line $\overline{A'A''}$ is parallel to $\overline{OS}$. By Thales' theorem, $\overline{AA''}/\overline{AO} = \overline{AA'}/\overline{AS}$; since the first quotient is $1/2$, we infer that A' (resp., B') is the midpoint of the segment $\overline{SA}$ (resp., $\overline{SB}$). Moreover, by applying Thales's theorem again we can see that $\overline{SA}/\overline{SA'} = \overline{SO}/\overline{SM}$; since the first quotient is equal to 2, we infer that O is the midpoint of $\overline{SM}$. Finally, since A' and B' are the midpoints of $\overline{SA}$ and $\overline{SB}$, it is easy to deduce that the segment $\overline{A'B'}$ is parallel to segment $\overline{AB}$. $\square$

All these deductions exposed in the form of a lemma were deducted by Archimedes more than 2200 years ago, so it is fair to summarize them in a theorem that bears his name.

THEOREM 8.1. *(Archimedes) The median to the base of an Archimedean triangle is parallel to the axis of the parabola, its midpoint also belongs to the parabola and the tangent at that point passes through the midpoints of the other two sides of the triangle.*

The great study of the parabola carried out by Archimedes did not end here: with the notation exposed so far, let us see how the Greek genius calculated the area J of the section of the parabola delimited by the Archimedean triangle ASB.

PROPOSITION 8.1. *The area J enclosed by the parabola section AOB (curve AOB and segment AB) is equal to two thirds of the area of the Archimedean triangle ASB.*

PROOF. The segments $\overline{A'B'}$, $\overline{OA}$ and $\overline{OB}$ divide the triangle ASB in four different sections (see figure 8.4): 1) the so-called "inner triangle" AOB, which is completely inside the parabola; 2) the so-called "outer triangle" $A'SB'$, which is completely outside the parabola although it contains a point of it; 3) and 4) the so-called "residual triangles" AOA' and BOB', which are also Archimedean triangles and the parabola passes through them.

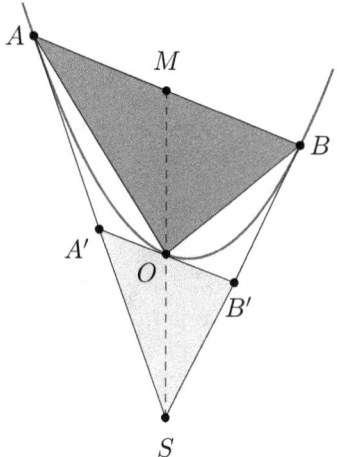

Figure 8.4

Since O is the midpoint of $\overline{SM}$, the first deduction we can make is that the area of the inner triangle is twice the area of the outer triangle, as both triangles have the same height and the inner triangle has a base twice as long as the outer triangle.

We must now apply an iterative process: in the same way that we have seen for the original Archimedean triangle, each of the two residual (Archimedean) triangles of figure 8.4 can be divided in four sections, always satisfying that the inner triangle has double the area of the outer one. Following this argument we can cover the entire surface of the initial Archimedean triangle ASB with inner and outer triangles, making the residuals triangles as small as we want. As for each inner triangle we have an outer triangle with an area that has a value of half the first one, we finally reach the conclusion that *"the area enclosed in a parabola section equals to two thirds of the area of its corresponding Archimedean triangle"*, result that Archimedes reached in antiquity. □

To complete the study, it is necessary to express this area in known geometric variables. Let us suppose that the equation of the parabola is $2py = x^2$ (p is the distance between the focus and the directrix axis) and we want to calculate the area enclosed between the parabola and the segment that joins its two points A and B (of coordinates (x_A, y_A) and (x_B, y_B) respectively) that are on the same quadrant of the coordinate plane.

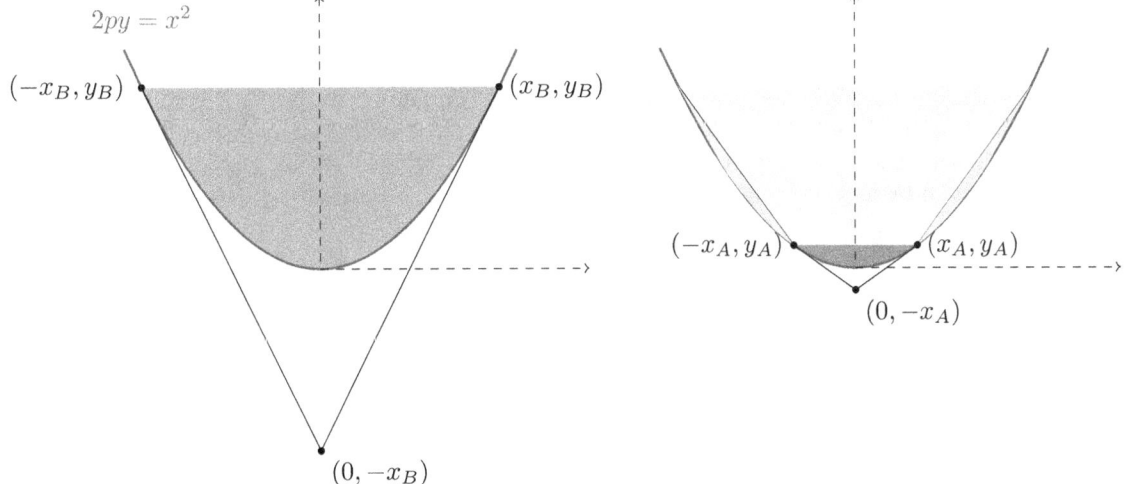

$$2py = x^2$$

$(-x_B, y_B)$ (x_B, y_B)

$(0, -x_B)$

$(-x_A, y_A)$ (x_A, y_A)

$(0, -x_A)$

Figure 8.5

As seen in figure 8.5, the area J that we want to calculate is that enclosed between the line that joins the points (x_A, y_A) and (x_B, y_B) and the segment of the parabola between these points (that area is identical to its symmetric with points $(-x_A, y_A)$ and $(-x_B, y_B)$). To achieve this, we will first calculate the auxiliary areas A_1 (that area enclosed between the line joining $(-x_B, y_B)$ and (x_B, y_B) and the segment of the parabola between these points), A_2 (the trapezoid formed by points $(-x_A, y_A)$, $(-x_B, y_B)$, (x_B, y_B) and (x_A, y_A)) and A_3 (that area enclosed between the line joining $(-x_A, y_A)$ and (x_A, y_A) and the segment of the parabola between these points). With the conclusions we have attained previously, it can be deduced that:

$$A_1 = \frac{2}{3} \cdot \left(\frac{(2x_B) \cdot (2y_B)}{2} \right)$$

$$A_2 = \frac{2x_A + 2x_B}{2} \cdot (y_B - y_A)$$

$$A_3 = \frac{2}{3} \cdot \left(\frac{(2x_A) \cdot (2y_A)}{2} \right)$$

To calculate A_1 (and A_3) we have applied Archimedes' conclusions ("*the area enclosed in a parabola section is equal to two thirds of the area of its corresponding Archimedean triangle*"), noting that the length of the base of the triangle is $2x_B$ and the height is $2y_B$ (remember that the vertex O is the midpoint between M and S). To calculate A_2 we have applied the formula of the area of a trapezoid with bases $2x_B$ and $2x_A$, and height $(y_B - y_A)$.

We know, by the equation of the parabola, that $2py_A = x_A^2$ and $2py_B = x_B^2$, so we finally obtain the following formula for area J:

$$J = \left(\frac{A_1 - A_2 - A_3}{2} \right) = \frac{2}{3} \cdot x_B \cdot y_B - \frac{x_A + x_B}{2} \cdot (y_B - y_A) - \frac{2}{3} \cdot x_A \cdot y_A =$$

$$= \frac{2}{3} \cdot x_B \cdot \frac{x_B^2}{2p} - \frac{x_A + x_B}{2} \cdot \frac{x_B^2 - x_A^2}{2p} - \frac{2}{3} \cdot x_A \cdot \frac{x_A^2}{2p} = \cdots = \frac{(x_B - x_A)^3}{12p}$$

Archimedes: **"The area of a parabolic section can be calculated as the cube of the projection of its end points onto the directrix line divided by 12 times the parameter of the parabola".**

FINAL REMARKS

- Obviously, this result can be found by calculating areas using integrals, but the Archimedean era was several centuries before this tool was available. From an objective point of view, Archimedes demonstrated with these results that he was a man advanced to his time and one of the best mathematicians in history.

- The interested reader can verify that this formula is also true if the chosen points of the parabola are not in the same quadrant of the coordinate plane.

Chapter 9

The parabola as an envelope

(Apollonius – 212 BC)

PROBLEM

We have an arbitrary angle. On each of the sides that form the angle we take $(n+1)$ points at a distance $k \cdot d$ $(k = 0, 1, ..., n)$ from the vertex. We join with a line the point of a side of the angle at distance $i \cdot d$ from the vertex and the point of the other side of the angle at distance $(n-i) \cdot d$ from the vertex (see figure 9.1). The problem is to prove that a parabola is the envelope of these $(n+1)$ lines.

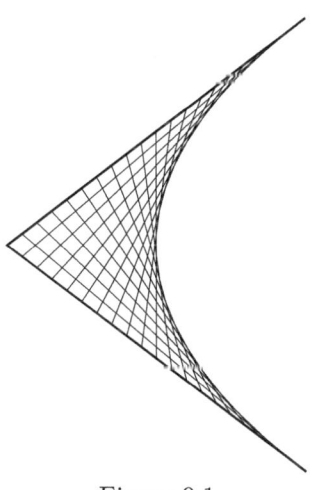

Figure 9.1

HISTORY

Apollonius of Perga (262 - 190 BC) was a Greek mathematician (although the ruins of Perga are now in Turkey), that made history by brilliantly contributing to the study of conics. In fact, he proposed the names of parabola, hyperbola and ellipse that are now accepted worldwide.

Twenty-five years younger than Archimedes, Apollonius must have known his work and he improved the study of the parabola that we have seen in the previous problem by providing the theorem that bears his name and the curious property that occupies us in this problem.

For Greek mathematicians, finding the figure that turns out to be the (tangent) frontier of the succession of a family of curves was a challenge. That border was later called the **envelope** of the family of curves and this is the first known example that contains a bit of difficulty. In the problem entitled "The astroid" we will see another beautiful problem related to envelopes and we will give an idea of a generic method to find them. However, on this occasion we will present the original solution of Apollonius, endowed with enormous elegance.

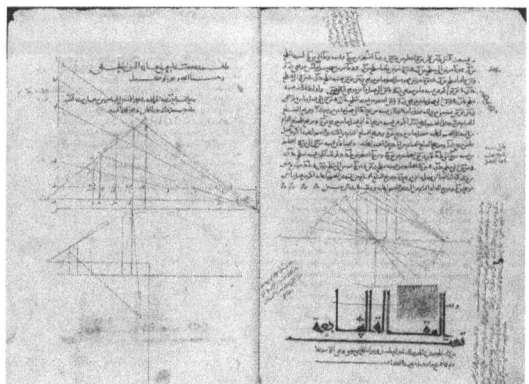

Fragments of the book "Conics" of Apollonius
Arab translation – 9th century

SOLUTION

To conclude that the desired envelope is a parabola we will demonstrate the so-called theorem of Apollonius and we will also use results of the previous problem of Archimedes.

Let A and B be two points of a parabola that are symmetric with respect to the axis (figure 9.2), S be the intersection point of the tangents to the parabola at A and B (by symmetry, S lies in the axis of the parabola) and O be a point of the parabola whose tangent cuts segment $\overline{AS}$ at point P and segment $\overline{BS}$ at point Q. Let D be the directrix of the parabola and let a, b, c, d, p, q be the distances $\overline{AP}$, $\overline{BQ}$, $\overline{PS}$, $\overline{QS}$, $\overline{PO}$ and $\overline{OQ}$, respectively. Finally, let a', b', c', d', p', q' be the distances of the projections of the previous segments onto D.

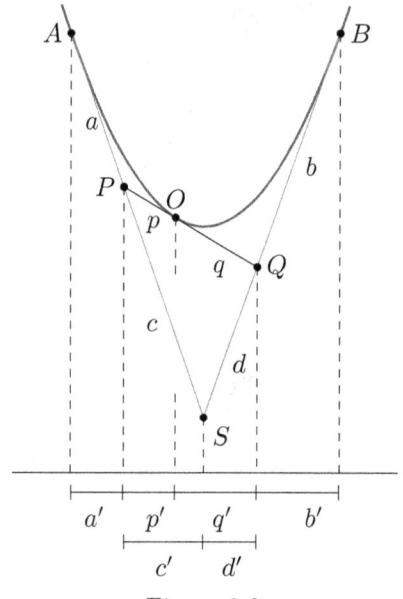

Figure 9.2

THEOREM 9.1. *(Apollonius) With the notation fixed above, it occurs that:*

$$\frac{c}{a} = \frac{q}{p} = \frac{b}{d}$$

PROOF. As we saw in Problem 8, an Archimedean triangle fulfills the property that its median to the base is parallel to the axis of the parabola. In figure 9.2 there are three Archimedean triangles (APO, ASB and BQO), so we can deduce, for each of them, that the length of the projection of the two sides that are not the base is the same (applying Thales' theorem). For the triangle APO

that means that $a' = p'$ (9.1); for the triangle ASB means that $a' + c' = b' + d'$ (9.2); and for the triangle BQO means that $b' = q'$ (9.3). Finally, by construction, it is clear that $p' + q' = c' + d'$ (9.4).

If we substitute equations (9.1) and (9.3) in equation (9.4) we find $a' + b' = c' + d'$ (9.5). If we now subtract equations (9.5) and (9.2) we see that $b' - c' = c' - b' \Rightarrow b' = c'$ (9.6), from which we also infer that $a' = d'$ (9.7) by substituting (9.6) in (9.2), for example.

Then:

- $c/a = c'/a' = b'/a'$ (first equality by Thales, second equality by (9.6))
- $q/p = q'/p' = b'/a'$ (first equality by Thales, second equality by (9.3) and (9.1))
- $b/d = b'/d' = b'/a'$ (first equality by Thales, second equality by (9.7))

Therefore, it follows that $c/a = q/p = b/d$. $\square$

The reciprocal of the theorem is also true:

COROLLARY 9.1. *A segment $\overline{PQ}$ with its endpoints in sides $\overline{SA}$ and $\overline{SB}$ that fulfills the equality $(\overline{SP}/\overline{PA}) = (\overline{QB}/\overline{SQ})$ is tangent to the parabola that passes trough A and B and has the lines $\overline{SA}$ and $\overline{SB}$ as tangents.*

PROOF. For any point P of the segment $\overline{SA}$, the quotient $\overline{SP}/\overline{PA}$ has a different value (if we think of P starting from S and ending in A, the value of $\overline{SP}$ increases while the value of $\overline{PA}$ decreases, so the value of $\overline{SP}/\overline{PA}$ is always increasing).

Moreover, we always have a tangent to the parabola that cuts the segment $\overline{AS}$ at a chosen point, since the tangents are traversing continuously the plane from $\overline{AS}$ (tangent at A) up to $\overline{SB}$ (tangent at B).

Using both reasonings, we conclude that if we have a point P and a point Q that fulfill $(\overline{SP}/\overline{PA}) = (\overline{QB}/\overline{SQ})$, then necessarily those P and Q are the same points that come from the only tangent that meets $\overline{SA}$ in P. In other words, there cannot be points P and Q that fulfill $(\overline{SP}/\overline{PA}) = (\overline{QB}/\overline{SQ})$ different from the points of the tangent that we know exists and fulfills the same equality by the theorem of Apollonius. $\square$

PROPOSITION 9.1. *The parabola that passes through A and B and has tangents $\overline{AS}$ and $\overline{BS}$ is the desired envelope.*

PROOF. We divide the segment AS (and BS) in n parts of equal length d and let us see one of the lines of the statement, for example, the line that starts at the point P of segment $\overline{AS}$ that is at distance $m \cdot d$ from S and at distance $(n - m) \cdot d$ from A. For the conditions of the problem, the point Q of the segment $\overline{BS}$ where the line ends is at distance $(n - m) \cdot d$ from S and at distance $m \cdot d$ from B (see figure 9.3).

Then, for that line it is fulfilled that:

$$\frac{\overline{SP}}{\overline{PA}} = \frac{\overline{QB}}{\overline{SQ}} = \frac{m \cdot d}{(n - m) \cdot d}$$

By the previous corollary, that means that the parabola described is tangent to this line. As this happens for all possible lines, the parabola is necessarily the desired envelope. $\square$

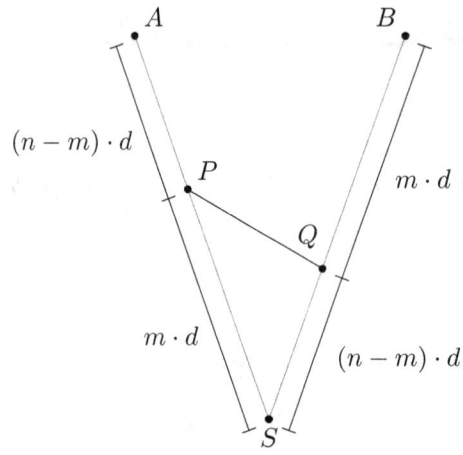

Figure 9.3

FINAL REMARKS

Obviously, in figure 9.3 we can also find the point O in the segment PQ that is tangent to the parabola, since it must be fulfilled by applying the Apollonius theorem:

$$\frac{\overline{SP}}{\overline{PA}} = \frac{\overline{QB}}{\overline{SQ}} = \frac{m \cdot d}{(n-m) \cdot d} = \frac{\overline{OQ}}{\overline{OP}}$$

That is, we can go calculating points of tangency of the parabola by applying the theorem. In general, any four points are necessary to draw a parabola, but in this case, as points A and B are symmetric with respect to the axis, it is only necessary another point.

Chapter 10

Change of direction of a planet

(Ptolemy – 145)

PROBLEM

To determine when a planet, as seen from Earth, will change its direction with respect to distant and fixed stars.

HISTORY

It is well-known that humanity placed Earth as the center of the Universe (geocentric theory) for centuries and tried to explain the movement of the rest of celestial bodies from that apparent truth.

However, the first problem of the theory was that anyone could observe with the naked eye the change of direction of the planets in the sky. Indeed, if Earth is the center of the universe and the rest of the planets revolve in circular orbits around it, how could it be explained that Venus moved during many days in a certain direction (with respect to the distant and fixed stars) and, suddenly, changed to the opposite direction?

Ptolemy devised a theory, based on Apollonius's previous work, according to which any planet revolved in small circles around a center, with this center being the point that followed a circular orbit with respect to Earth, as we can see in figure 10.1. With this trick, the change of direction of the planets could apparently be explained without disturbing the importance of Earth as the center of the Universe.

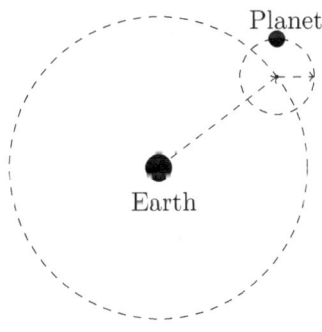

Figure 10.1

This theory, called epicyclic theory, was defended by the majority as true until the sixteenth century (and beyond by some of its most fervent followers), when Copernicus and Kepler made it obsolete by mathematically demonstrating the accuracy of the heliocentric model. Even before the validity of the epicyclic theory was in question, since more and more auxiliary circles were needed to try to explain the movements of celestial bodies, whose calculation improved with the greater precision of the measurements of observations.

Centuries later, many erroneous theories (but nevertheless tenaciously defended by someone in spite of the experimental observations against them) are ridiculed when compared with the epicyclic theory, an example of centuries of human obstinacy and stubbornness.

Ptolemy inspired by Urania, muse of the Astronomy
Illustration of "Margarita Philosophica" – Gregor Reisch (1508)

In the proposed solution to the problem (based, obviously, on the heliocentric model) the reason for the change of direction of a planet seen from Earth will be explained. We will put forward for this the simplifications that the orbits of the planets are circular around the Sun and that they lie all in the same plane.

SOLUTION

Let E be Earth and suppose that it orbits circularly around the Sun at constant angular velocity w_E. Let us put the Sun at the center of our origin of rectangular coordinates and Earth on the axis X at the moment we take as the initial time of the problem. Finally, let R_E be the Earth - Sun distance (constant, by our circular orbit assumption).

Now let P be another planet, which orbits around the Sun at constant angular velocity w_P and **in the same plane** as the Earth's orbit. Let R_P be the distance, also constant, Planet - Sun (suppose that $R_P < R_E$, although the solution to the problem would be identical otherwise) and suppose that the segment Planet - Sun forms an angle α with the positive side of the X axis at the initial time of the problem. Everything exposed so far can be seen in figure 10.2.

The equation of the movement of E as a function of time is $\vec{E}(t) = (R_E \cos w_E t, R_E \sin w_E t)$, while the equation of P turns out to be $\vec{P}(t) = (R_P \cos(w_P t + \alpha), R_P \sin(w_P t + \alpha))$ (note that at the instant $t = 0$ we have $\vec{P}(0) = (R_P \cos \alpha, R_P \sin \alpha)$, as we stated before).

Therefore, the vector that joins E and P follows the equation $\vec{L}(t) = (R_P \cos(w_P t + \alpha) - R_E \cos w_E t, R_P \sin(w_P t + \alpha) - R_E \sin w_E t)$, which forms an angle β (not constant in time) with the positive half axis of X (see again figure 10.2).

This vector shows the direction from which an observer on Earth sees the planet P. Furthermore, distant stars do not vary practically their position seen from Earth, which is the reason why an observer would see the planet move with respect to them in the same way that the angle β varies with respect to a fixed direction (see figure 10.3)

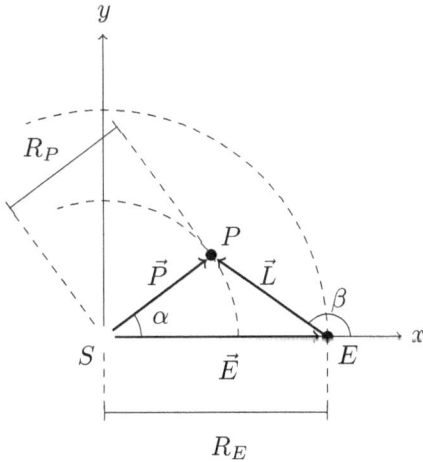

Figure 10.2

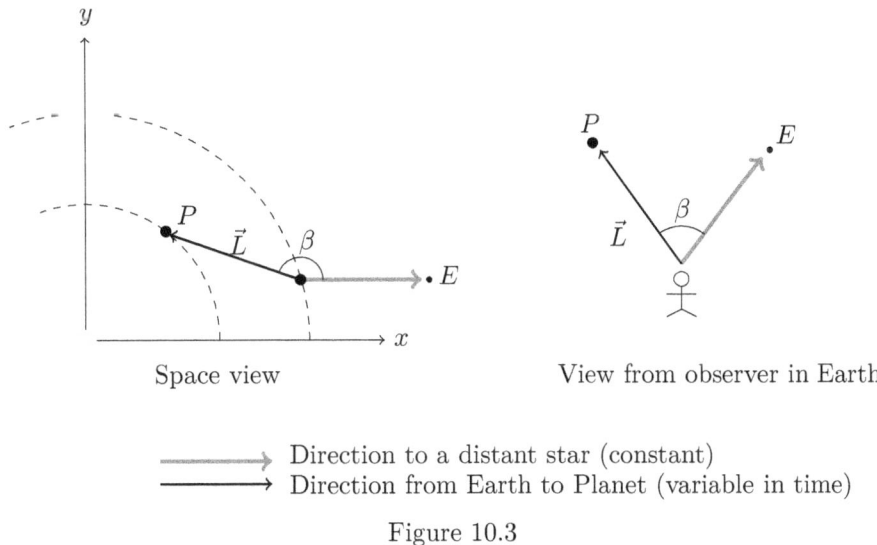

Space view   View from observer in Earth

———→ Direction to a distant star (constant)
———→ Direction from Earth to Planet (variable in time)

Figure 10.3

That is, we can consider that a fixed observation point in the sky is in a fixed direction of our coordinate system. For example, we can think of a very distant star that always is in the same direction of the positive axis of X (the argument does not change if we think in another direction). The assumption that it is fixed does not give an error if we consider a star that is several light-years away from Earth.

An observer in Earth can see that the angle between β and the positive axis of X varies in time and, therefore, can increase or decrease. Precisely when changes from increasing to decreasing (or vice versa) it is when we see in heaven the "change of direction" of the planet. In other words, we must observe the function $\beta(t)$ and notice when it goes from increase to decrease, that is, to look for its extremes by forcing its derivative to be equal to 0, $\beta'(t) = 0$.

To do this, let us first calculate the function corresponding to $f(t) = \tan[\beta(t)]$:

$$f(t) = \tan[\beta(t)] = \frac{R_P \sin(w_P t + \alpha) - R_E \sin w_E t}{R_P \cos(w_P t + \alpha) - R_E \cos w_E t}$$

It is not necessary now to isolate $\beta(t)$ and to derive it later. If we want to find the values of $\beta'(t) = 0$, we can do it analogously looking for those that fulfill $f'(t) = 0$, since by the chain rule in derivation, $f'(t) = \sec^2[\beta(t)] \cdot \beta'(t)$ (multiplying by a value does not vary the search for zeros).

Therefore, let us look for the solutions of $f'(t) = 0$:

$$f'(t) = \tan[\beta(t)] = \frac{(R_P w_P \cos(w_P t + \alpha) - R_E w_E \cos w_E t) \cdot (R_P \cos(w_P t + \alpha) - R_E \cos w_E t)}{(R_P \cos(w_P t + \alpha) - R_E \cos w_E t)^2} -$$

$$- \frac{(R_P \sin(w_P t + \alpha) - R_E \sin w_E t) \cdot (-R_P w_P \sin(w_P t + \alpha) + R_E \sin w_E t)}{(R_P \cos(w_P t + \alpha) - R_E \cos w_E t)^2} = 0 \quad \Rightarrow$$

$$\Rightarrow \quad \left[R_P^2 w_P \cos^2(w_P t + \alpha) - R_P R_E (w_P + w_E) \cos(w_P t + \alpha) \cos(w_E t) + R_E^2 w_E \cos^2(w_E t) \right] +$$

$$+ \left[R_P^2 w_P \sin^2(w_P t + \alpha) - R_P R_E (w_P + w_E) \sin(w_P t + \alpha) \sin(w_E t) + R_E^2 w_E \sin^2(w_E t) \right] = 0$$

By applying the trigonometric theorems $\cos(ab) = \cos(a)\cos(b) - \sin(a)\sin(b)$ and $\cos^2 a + \sin^2 a = 1$:

$$R_P^2 w_P - R_P R_E (w_P + w_E) \cos((w_P - w_E)t + \alpha) + R_E^2 w_E = 0 \quad \Rightarrow$$

(10.1) $$\quad \Rightarrow \quad \cos((w_P - w_E)t + \alpha) = \frac{R_P^2 w_P + R_E^2 w_E}{R_P R_E (w_P + w_E)}$$

That is, when t is a value that makes the equation (10.1) true, then $f'(t) = 0$, which implies that $\beta'(t) = 0$, which means in turn (as we explained before) an "apparent" change of direction of P seen from Earth with respect to a fixed direction in the sky.

Taking advantage of one of Kepler's laws, we can simplify the right part of the expression (10.1). According to that law, the radii of the orbits and the angular velocities of the planets of the Solar System fulfill a curious property: *"The ratio between the cubes of the orbital radii is equal to the inverse of the ratio between the squares of the angular velocities"*, that is,

(10.2) $$\frac{w_E^2}{w_P^2} = \frac{R_P^3}{R_E^3}$$

Actually, what we have called "a curious property" can be deduced from Newton's gravitational theory. Now, if we define $r_P = \sqrt{R_P}$ and $r_E = \sqrt{R_E}$, then equation (10.2) can be written as:

(10.3) $$\frac{w_E}{w_P} = \frac{r_P^3}{r_E^3}$$

The right-hand part of equation (10.1) can be simplified by applying equation (10.3):

$$\frac{R_P^2 w_P + R_E^2 w_E}{R_P R_E (w_P + w_E)} = \frac{r_P^4 (w_P/w_E) + r_E^4}{r_P^2 r_E^2 (w_P/w_E + 1)} = \frac{r_P^4 (r_E^3/r_P^3) + r_E^4}{r_P^2 r_E^2 (r_E^3/r_P^3 + 1)} = \frac{r_P^4 r_E^3 + r_E^4 r_P^3}{r_P^2 r_E^2 (r_E^3 + r_P^3)} =$$

$$= \frac{r_P^2 r_E + r_E^2 r_P}{r_E^3 + r_P^3} = \frac{(r_E + r_P) \cdot r_P r_E}{(r_E + r_P) \cdot (r_E^2 - r_P r_E + r_P^2)} = \frac{r_P r_E}{r_E^2 - r_P r_E + r_P^2} = \frac{\sqrt{R_P R_E}}{R_E - \sqrt{R_P R_E} + R_P}$$

Finally, equation (10.1) that gives time solutions when planet P changes of direction, is as follows:

$$(10.4) \qquad \cos((w_P - w_E)t + \alpha) = \frac{\sqrt{R_P R_E}}{R_E - \sqrt{R_P R_E} + R_P}$$

FINAL REMARKS

Real example with data from the Earth's and Venus' orbits

Earth's orbit can be approximated by a circle of radius $R_E = 149$ million km and angular velocity $w_E = 0.9856$ degrees/day (completing $360°$ in 365 days). Venus' orbit can be seen as another circumference with $R_P = 107.5$ million km and $w_P = 1,602$ degrees/day (the Venus "year" lasts $360/1,602 = 224.7$ terrestrial days).

If we assume that on a particular day the planets are perfectly aligned and with the Sun in between ($\alpha = 180°$), with the formula (10.4) we can calculate how long it will take for the first apparent change of Venus direction seen by a terrestrial observer:

$$\cos(0.6164t + 180) = \frac{\sqrt{107.5 \cdot 149}}{149 - \sqrt{107.5 \cdot 149} + 107.5} = 0.974$$

The first solution of the above equation occurs when $t = 271$. That is, Venus will change its direction in the sky 271 "terrestrial" days after the initial position indicated.

Transit of Venus

In fact, it is not true that the Earth and Venus' orbits are in the same plane. If so, from time to time (about every eight years) Venus would be in the middle of Earth and the Sun, causing a tiny eclipse (Venus would be a small spot in the immense Sun) that is usually called a "transit of Venus".

But the planes of the orbits differ by about $3.4°$, which causes that there is no alignment every eight years, but only sporadically. However, two separate transits occur in a period of eight years (the small deviation produces that after eight years of a transit, Venus still does not deviate enough so that in the next alignment a new eclipse does not occur), taking longer then 100 years to go back to produce another. Many people, therefore, will not witness a transit of Venus throughout their lives, but some of us have been able to enjoy the last two, which occurred in June 2004 and June 2012; the next one will occur in December 2117

Chapter 11

Pell's equation

(Brahmagupta – 628)

PROBLEM

To find the integer solutions of the equation $x^2 - d \cdot y^2 = 1$, where d is a positive integer that is not a square.

HISTORY

The Indian and Greek mathematicians of the 4th century BC were already interested in this equation and its connection to fractions of integers that approximated square roots. Later, the great Indian mathematician Brahmagupta found a partial solution (valid only for certain values of d) in his masterpiece "The doctrine of Brahma correctly established" (628 d.C.).

Apparently, nobody in Europe knew that book, so in 1567 (a thousand years later!) Pierre de Fermat proposed this equation as a problem to the most prestigious mathematicians of the continent; in the exchange of letters that followed, the Englishman Lord William Brounker (1620 - 1684) found a general method to solve it.

Brahmagupta (598 – 660)

However, Leonhard Euler mistook his work with that of John Pell (1611 - 1685), who solely explained Brounker's solution in a book. Unfortunately for Brahmagupta and Brounker, Euler's reputation was enormous and the name "Pell's equation" has survived to this day, despite some laudable attempts to rename it in honor of the true discoverers.

The beautiful solution that we present below, based on continued fractions and the study of the properties of the numbers of the form $P + Q\sqrt{d}$, was developed by the Italian - French mathematician Joseph - Louis Lagrange (1736 - 1813), who was appointed Count by Napoleon for his innumerable contributions to science and buried in the Pantheon in Paris.

SOLUTION

First of all, it should be noted that the imposition of d not being a square is due to the fact that if it were $(d = c^2)$, we would have that the original equation could be written as $x^2 - d \cdot y^2 = 1 \Rightarrow (x - cy) \cdot (x + cy) = 1$, which would force, as all numbers are integers, that $(x - cy) = (x + cy) = \pm 1$ and, in those cases, only trivial solutions $(1, 0)$ and $(-1, 0)$ exist. Therefore, if we want the problem to be interesting, it is necessary to impose the condition that d is not a square and, therefore, that $\sqrt{d}$ is an irrational number.

Approach to the solution

The first mathematicians who tried to solve this equation came to the following conclusion: suppose we have a set of fractions p_n/q_n of positive integers that approximate to the irrational value $\sqrt{d}$, that is, $p_n/q_n \approx \sqrt{d}$. Then, we will have $p_n^2/q_n^2 \approx d$ and therefore that $p_n^2 - d \cdot q_n^2 \approx 0$. Since p_n, q_n are positive integers, the exact (not approximate) value of $p_n^2 - d \cdot q_n^2$ will also be an integer "close" to 0. Then it is possible that for some pair p_n, q_n, that integer is equal to 1 and we find a solution to Pell's equation.

For this reason, they began looking for fractions that approximate the irrational numbers of the form $\sqrt{d}$ and the theory of continued fractions was born, some of whose results we will list and demonstrate next. Obviously, the above reasoning could not lead to any result (it may never be possible to get $p_n^2 - d \cdot q_n^2$ equal to 1, even though it is "close" to 0), but it seemed to be a good starting point.

Fortunately, as we will see, the theory of continued fractions does find solutions to Pell's equation.

Continued fractions

DEFINITION 11.1. *Let α be an irrational number. We define a **continued fraction** of α as the series of integers $(a_0, a_1, a_2, ...)$, called **coefficients**, which are found recursively with the help of the real numbers $(r_0, r_1, r_2, ...)$, called **residuals**, as follows:*

n	r_n	a_n
0	$r_0 = \alpha$	$a_0 = \lfloor r_0 \rfloor$
1	$r_1 = \dfrac{1}{r_0 - \lfloor r_0 \rfloor}$	$a_1 = \lfloor r_1 \rfloor$
	...	
n	$r_n = \dfrac{1}{r_{n-1} - \lfloor r_{n-1} \rfloor}$	$a_n = \lfloor r_n \rfloor$

where $\lfloor x \rfloor$ is the greatest integer that is less or equal to x.

Example. For $\alpha = \sqrt{7}$ we find the values:

n	r_n	a_n
0	$r_0 = \sqrt{7} = 2.645...$	$a_0 = \lfloor \sqrt{7} \rfloor = 2$
1	$r_1 = \dfrac{1}{\sqrt{7} - 2} = 1.5485...$	$a_1 = \lfloor 1.5485... \rfloor = 1$
2	$r_2 = \dfrac{1}{1.5485... - 1} = 1.8228...$	$a_2 = \lfloor 1.8228... \rfloor = 1$
	...	

Therefore, the continued fraction of $\sqrt{7}$ starts with $(2, 1, 1, ...)$.

Actually, for irrational numbers that are square roots of integers, you can (and should) work with exact r_n values. In this case you can check that the residuals are all of the form $(m + \sqrt{7})/k$ (where m, k are positive integers):

$$r_0 = \frac{0 + \sqrt{7}}{1}, \qquad r_1 = \frac{2 + \sqrt{7}}{3}, \qquad r_2 = \frac{1 + \sqrt{7}}{2}, \qquad \cdots$$

Why looking for these numbers in this way? From the definition you can deduce (with a little patience) the following equalities:

(11.1)
$$\alpha = r_0, \qquad \alpha = a_0 + \frac{1}{r_1}, \qquad \alpha = a_0 + \frac{1}{a_1 + \frac{1}{r_2}}, \qquad \cdots$$

If the value of r_n in the previous equations is replaced by a_n we convert the equalities into approximations:

(11.2)
$$\alpha \approx a_0, \qquad \alpha \approx a_0 + \frac{1}{a_1}, \qquad \alpha \approx a_0 + \frac{1}{a_1 + \frac{1}{a_2}}, \qquad \cdots$$

So what we propose is a way to approximate irrational numbers with a series of integer fractions. The more values of the series we calculate the closer we will get to the original number, hence the name of a continued fraction of an irrational.

Each of the approximations we have seen in (11.2), which we call S_n, is better than the previous one. In our example, $\alpha = \sqrt{7} = 2.645751...$:

$$S_0 = a_0 = 2$$
$$S_1 = a_0 + \frac{1}{a_1} = 2 + \frac{1}{1} = 3$$
$$S_2 = a_0 + \frac{1}{a_1 + \frac{1}{a_2}} = 2 + \frac{1}{1 + \frac{1}{1}} = \frac{5}{2} = 2.5$$

The reader can verify, with patience, that $S_3 = 8/3 = 2.666666...$, $S_4 = 37/14 = 2.642857...$, $S_5 = 45/17 = 2.647058...$ and $S_6 = 82/31 = 2.645161...$, for example, with each step getting closer to the original value.

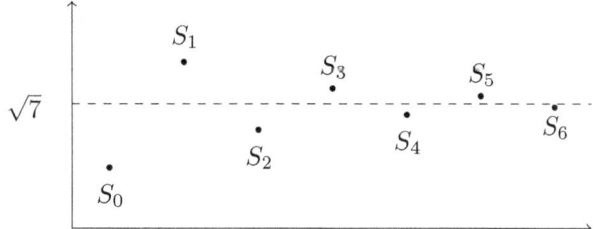

Figure 11.1

61

The idea of this book is not to prove all the following results, but the reader can do it with the help of some books on continued fractions:

- The value of S_n is closer than S_{n-1} to the value of α.
- The values $S_0, S_2, ..., S_{2k}$ are increasing but always smaller than α, while $S_1, S_3, ..., S_{2k+1}$ are decreasing but always greater than α.
- The series S_n has a limit and its value is α.
- The residuals r_n are always irrational numbers (we are assuming α is irrational) and, except maybe r_0, the rest of them are greater than one, which forces that coefficients a_n ($n > 1$) are natural numbers.

By construction, the value of S_n is a rational number that can be written as a fraction p_n/q_n (both integers, with no common divisors), and these fractions (called **convergents**) will be the candidates to fulfill $p_n^2 - d \cdot q_n^2 = 1$ when we compute them for the value of $\alpha = \sqrt{d}$.

Now, for the calculation of p_n/q_n we need an efficient method that is better than summing so many fractions as in (11.2), something very cumbersome. We state the following lemma:

LEMMA 11.1. *With the notation used so far and assuming that we have already calculated the coefficients $(a_0, a_1, a_2, ...)$, it can be deduced that, for $n \geq 2$:*

(11.3)
$$\begin{cases} p_n = a_n p_{n-1} + p_{n-2} \\ q_n = a_n q_{n-1} + q_{n-2} \end{cases}$$

PROOF. By induction on n. We leave it to the interested reader, who can easily consult it. $\square$

The recursive formulas (11.3) yield an easy method to calculate the values of p_n/q_n. In the following table we assume that, for $\alpha = \sqrt{7}$, we know all the values a_n (we have calculated them as we did previously) and that we have calculated those of p_0, q_0, p_1, q_1 from the definition. For the rest of values, we only have to apply (11.3): for example, to find the value $p_4 = 37$ in the table below, we have calculated $a_4 \cdot p_3 + p_2 = 4 \cdot 8 + 5$ (all values in the table, in previous lines).

n	a_n	p_n	q_n	p_n/q_n
0	2	2	1	$2/1 = 2$
1	1	3	1	$3/1 = 3$
2	1	5	2	$5/2 = 2.5$
3	1	8	3	$8/3 = 2.66666...$
4	4	37	14	$37/14 = 2.64285...$
5	1	45	17	$45/17 = 2.64705...$
6	1	82	31	$82/31 = 2.64516...$
7	1	127	48	$127/48 = 2.64583...$
...				$\sqrt{7} = 2.64575...$

A small change in lemma 11.1 leads us to another interesting result:

COROLLARY 11.1. *For $n \geq 1$:*
$$\alpha = \frac{r_{n+1} p_n + p_{n-1}}{r_{n+1} q_n + q_{n-1}}$$

PROOF. If we look the equations (11.1) and (11.2) we wrote when we explained the definition of continued fraction, we can see that (11.1) are equalities where α and the coefficients

$(a_0, \cdots, a_n, r_{n+1})$ appear; instead, in (11.2) the equalities are the same but with fractions p_n/q_n (which we define later) and coefficients $(a_0, \cdots, a_n, a_{n+1})$.

Therefore, the proof of lemma 11.1 can be repeated in the same way but using α instead of p_n/q_n and using $(a_0, \cdots, a_n, r_{n+1})$ instead of $(a_0, \cdots, a_n, a_{n+1})$, so we can write:

$$\alpha = \frac{r_{n+1}p_n + p_{n-1}}{r_{n+1}q_n + q_{n-1}} \qquad \text{instead of} \qquad \frac{p_n}{q_n} = \frac{a_{n+1}p_n + p_{n-1}}{a_{n+1}q_n + q_{n-1}}$$

which is what we want to prove. $\qquad \qquad \square$

Lemma 11.1 is not only used in order to efficiently calculate the values of p_n, q_n; the result will serve us later. For example, for the following lemma:

LEMMA 11.2. *If $n \geq 1$, then $p_n q_{n-1} - p_{n-1} q_n = (-1)^{n-1}$.*

PROOF. By induction on n. For $n = 1$, the statement is true:

$$p_1 q_0 - p_0 q_1 = (a_0 a_1 + 1) \cdot (1) - (a_0) \cdot (a_1) = 1$$

Now, if the result is true for n, then for $n + 1$:

$$p_{n+1} q_n - p_n q_{n-1} \stackrel{(a)}{=} (a_{n+1} p_n + p_{n-1}) \cdot q_n - p_n \cdot (a_{n+1} q_n + q_{n-1}) =$$

$$= p_{n-1} q_n - p_n q_{n-1} \stackrel{(b)}{=} -(-1)^{n-1} = (-1)^n$$

where we have applied lemma 11.1 in (a) and the hypothesis of induction for n in (b). $\qquad \square$

These results are true for any irrational number α, but now let us focus on our goal, when $\alpha = \sqrt{d}$.

PROPOSITION 11.1. *If $\alpha = \sqrt{d}$ then $r_i = \frac{m_i + \sqrt{d}}{k_i}$, where m_i, k_i are integers that fulfill the following recursive equations:*

$$m_{i+1} = a_i k_i - m_i \qquad \text{and} \qquad k_{i+1} = \frac{d - m_{i+1}^2}{k_i}$$

PROOF. When we calculated some residuals for $\sqrt{7}$ we already intuitively saw that they are all of the type $r_i = (m_i + \sqrt{7})/k_i$.

For the general case, let us prove it by induction. First, for r_1:

$$r_1 = \frac{1}{\sqrt{d} - a_0} = \frac{1}{\sqrt{d} - a_0} \cdot \frac{\sqrt{d} + a_0}{\sqrt{d} + a_0} = \frac{\sqrt{d} + a_0}{d - a_0^2}$$

where $m_1 = a_0 k_0 - m_0 = a_0 \cdot 1 - 0$ and $k_1 = (d - m_1^2)/k_0 = d - a_0^2$, as the proposition states. Now, if the identity is true for $r_n = (m_n + \sqrt{d})/k_n$, we have to prove it for r_{n+1}:

63

$$r_{n+1} = \cfrac{1}{\frac{m_n + \sqrt{d}}{k_n} - a_n} = \frac{k_n}{(m_n - a_n \cdot k_n) + \sqrt{d}} = \frac{k_n}{(m_n - a_n \cdot k_n) + \sqrt{d}} \cdot \frac{(m_n - a_n \cdot k_n) - \sqrt{d}}{(m_n - a_n \cdot k_n) - \sqrt{d}} =$$

$$= \frac{(a_n \cdot k_n - m_n) + \sqrt{d}}{\frac{d - (m_n - a_n \cdot k_n)^2}{k_n}} := \frac{m_{n+1} + \sqrt{d}}{k_{n+1}}$$

where we have defined $m_{n+1} = a_n k_n - m_n$ and $k_{n+1} = (d - m_{n+1}^2)/k_n$.

We only have to make sure that k_{n+1} is an integer, but that can also be proven by induction, since $k_{n+1} = (d - (m_n - a_n k_n)^2)/k_n$ and the denominator k_n divides the numerator because it can be written as $d - (m_n - a_n k_n)^2$, or $(dm_n^2) + k_n \cdot (2a_n m_n + a_n^2 k_n)$, the two addends being divisible by k_n (the first one by the induction's hypothesis, since $dm_n^2 = k_n \cdot k_{n+1}$, and the second one is trivial). $\qquad\square$

Now we face the core of the problem. We are going to prove that for $\alpha = \sqrt{d}$ a curious property is satisfied in the continued fraction: at a certain iteration, coefficients begin to repeat themselves periodically, so that it is fulfilled that, at some point, $a_j = a_i$ $(j > i)$ and, thereafter, $a_{j+n} = a_{i+n}$ ($\forall n$ positive integer). For example, for $\alpha = \sqrt{7}$ it is satisfied that the continued fraction has coefficients $(2, 1, 1, 1, 4, 1, 1, 1, 4, ...)$ with the sequence $\{1, 1, 1, 4\}$ indefinitely repeated. When that happens, we say that the continued fraction is **periodic**, with period $N = j - i$ (in our example, $N = 4$) and we write it $(a_0, ..., a_{i-1}, \overline{a_i, ..., a_{j-1}})$ (in our example, $(2, \overline{1, 1, 1, 4})$).

PROPOSITION 11.2. *The continued fraction of $\alpha = \sqrt{d}$ is periodic.*

PROOF. First we are going to prove that, with the notation of proposition 11.1, the following inequations hold:

$$|m_i| < \sqrt{d} \qquad \text{and} \qquad 0 < k_i < d \qquad\qquad (\forall i \geq 0)$$

We use induction again. For $i = 0$ the statement is true, because $|m_0| = 0 < \sqrt{d}$ and $0 < k_0 = 1 < d$. Suppose now that it is true for i and let us try to prove it for $i + 1$:

$$|m_{i+1}| = |a_i k_i - m_i| = \left| \left\lfloor \frac{m_i + \sqrt{d}}{k_i} \right\rfloor \cdot k_i - m_i \right| \overset{(a)}{<} |(m_i + \sqrt{d}) - m_i| = \sqrt{d}$$

$$k_{i+1} = \frac{d - m_{i+1}^2}{k_i} \overset{(b)}{<} \frac{d}{k_i} \overset{(c)}{<} d$$

Inequality (a) is true because $(m_i + \sqrt{d})$ and k_i are positive (both statements, by hypothesis of induction), (b) is true since $0 < dm_{n+1}^2 < d$ (again, because $(m_i + \sqrt{d})$ is positive) and (c) is true because k_i is positive. In addition to the two previous inequalities, it is easy to see that k_{i+1} is positive because it is a division between positive numbers (both $d - m_{i+1}^2$ and k_i are positive).

After this step, we now can deduce that the value $r_i = (m_i + \sqrt{d})/k_i$ can only take a finite number of possibilities, since both m_i and k_i are integers and have both upper and lower bounds. Therefore, the residues, at some point, must be repeated. For example $r_j = r_i$ and, by the definition of residues, that means that, from then on and because all residues are calculated as we have explained before, $(r_{j+n} = r_{i+n})$ and the same occurs for the coefficients of the continued fraction $(a_{j+n} = a_{i+n})$. The continued fraction is, therefore, periodic. $\qquad\square$

We have to prove one more result only, found by Lagrange in 1770, and that leaves, finally, a cleared path to find the solutions of Pell's equation.

DEFINITION 11.2. *We say that a number that can be written as $\beta = (P + \sqrt{d})/Q$ where P, Q are positive integers is a **irrational quadratic**. We say that the **conjugated** of β, denoted as β', is the value $\beta' = (P - \sqrt{d})/Q$. If $\beta > 1$ and $-1 < \beta' < 0$, then we say that β is a **reduced irrational quadratic**.*

THEOREM 11.1. *(Lagrange) A reduced irrationally quadratic has a pure periodic continued fraction $(\overline{a_0, ..., a_j})$, that is, its period begins at its first term.*

PROOF. Let β be a reduced irrationally quadratic that has $(a_0, ..., a_k, ...)$ as the coefficients of its continued fraction. Then, its residue r_1 is also an irrational quadratic, since:

- $r_1 = 1/(\beta - \lfloor \beta \rfloor)$, so $r_1 > 1$ because $0 < \beta - \lfloor \beta \rfloor < 1$ (an irrational number minus its integer part).
- Its conjugate is $r_1' = 1/(\beta' - \lfloor \beta \rfloor)$, so $-1 < r_1' < 0$ because $\beta' - \lfloor \beta \rfloor < -1$ (a negative number minus a positive integer).

So, we have started with a reduced irrational quadratic β with coefficients $(a_0, ..., a_k, ...)$ and we have obtained another reduced irrational quadratic (whose first residue is r_1) with coefficients $(a_1, ..., a_k, ...)$, since we have advanced one step in the algorithm of the definition of continued fractions. But we can continue applying this reasoning to infer the same for $r_2, r_3, ...$ and all the residues, which must therefore also be reduced irrational quadratics.

We have by definition that $r_{k+1} = 1/(r_k - a_k)$, which is also true for its conjugates $r_{k+1}' = 1/(r_k' - a_k)$ or, which is the same, $a_k = (-1/r_{k+1}') + r_k'$. But in the previous paragraph we saw that $-1 < r_k' < 0$ (a reduced irrational quadratic), which helps us determine that $a_k = \lfloor -1/r_{k+1}' \rfloor$ (since $-1/r_{k+1}'$ is a positive irrational number greater than 1 to which we subtract a negative number greater than -1 to find an integer). That is:

$$(11.4) \qquad a_k = \left\lfloor \frac{-1}{r_{k+1}'} \right\rfloor \qquad \forall k \geq 0$$

Now, proposition 11.2 shows us that a reduced irrational quadratic has a periodic continued fraction (actually we saw it for $\alpha = \sqrt{d}$, but the proof is analogous), so it is true that, for a certain $j > k \geq 1$, $r_{j+n} = r_{k+n}$ $\forall n$ (and their corresponding coefficients). So, from what is found in (11.4), we have:

$$a_{j+n-1} = \left\lfloor \frac{-1}{r_{j+n}'} \right\rfloor \text{ and } a_{k+n-1} = \left\lfloor \frac{-1}{r_{k+n}'} \right\rfloor$$

If $r_{j+n} = r_{k+n}$ then $r_{j+n}' = r_{k+n}'$ and, therefore, $a_{j+n-1} = a_{k+n-1}$, from which follows $r_{j+n-1} = r_{k+n-1}$ (since, by definition, $r_{j+n-1} = a_{j+n-1} + 1/r_{j+n}$ and $r_{k+n-1} = a_{k+n-1} + 1/r_{k+n}$).

That is, we were able to "go back" one step in the coefficients equality $r_{j+n} = r_{k+n}$ and prove $r_{j+n-1} = r_{k+n-1}$. This can be repeated indefinitely until we reach $r_{j-k} = r_0$, so the period begins at the first term. □

The reader can check all these properties in an example of reduced irrational quadratic such as $2 + \sqrt{7}$.

COROLLARY 11.2. *Coefficients of the continued fraction of $\alpha = \sqrt{d}$ can be written as $(a_0, \overline{a_1, ..., a_N})$ (that is, the period always starts at the second coefficient), where $k_N = 1$.*

PROOF. $\alpha = \sqrt{d}$ is not a reduced irrational quadratic ($\alpha > 1$ but $\alpha' = -\sqrt{d} < -1$), but $\beta = \lfloor\sqrt{d}\rfloor + \sqrt{d} = a_0 + \sqrt{d}$ is a a reduced irrational quadratic ($\beta > 1$ and $-1 < \beta' = a_0 - \sqrt{d} < 0$). So, by the previous theorem, β has a pure continued fraction and, in this case, it starts with $2a_0$. We can write it as $(\overline{2a_0, a_1, ..., a_M})$ and, therefore, $\alpha = \beta - a_0$ has a periodic continued fraction $(a_0, \overline{a_1, ..., a_M, 2a_0})$ (the first coefficient has changed because we have subtracted a_0 (an integer) to β, but after that all the residues are the same), which we can write as $(a_0, \overline{a_1, ..., a_N})$.

In order to prove the second statement, we start with $a_N = 2a_0$ (as we have seen in the previous paragraph), which means that $r_N = (a_0 + \sqrt{d})/1$ (because $r_N = (m_N + \sqrt{d})/k_N$ and in the proof of proposition 11.2 we saw that $|m_N| < \sqrt{d}$, so that means $m_N = a_0$ and $k_N = 1$ if we want that $a_N = \lfloor r_N \rfloor$ reaches the value of $2a_0$). So, $k_N = 1$. $\qquad\square$

Solutions to Pell's equation

All previous work leads us to the following theorem:

THEOREM 11.2. *Let $\alpha = \sqrt{d}$ be a number with the residues of its continued fraction equal to $r_n = (m_n + \sqrt{d})/k_n$ and its convergents equal to p_n/q_n. Then, $p_{n-1}^2 - d \cdot q_{n-1}^2 = (-1)^n \cdot k_n \qquad \forall n \geq 1$.*

PROOF. In corollary 11.1 we prove that:

$$\alpha = \frac{r_n p_{n-1} + p_{n-2}}{r_n q_{n-1} + q_{n-2}} \qquad (\forall n > 1)$$

Therefore:

$$\sqrt{d} = \frac{\left(\frac{m_n + \sqrt{d}}{k_n}\right) \cdot p_{n-1} + p_{n-2}}{\left(\frac{m_n + \sqrt{d}}{k_n}\right) \cdot q_{n-1} + q_{n-2}} = \frac{(m_n + \sqrt{d}) \cdot p_{n-1} + k_n \cdot p_{n-2}}{(m_n + \sqrt{d}) \cdot q_{n-1} + k_n \cdot q_{n-2}} \qquad \Rightarrow$$

$$\Rightarrow \qquad \sqrt{d} \cdot \left[(m_n + \sqrt{d}) \cdot q_{n-1} + k_n \cdot q_{n-2} \right] = (m_n + \sqrt{d}) \cdot p_{n-1} + k_n \cdot p_{n-2} \qquad \Rightarrow$$

$$\Rightarrow \qquad (d \cdot q_{n-1}) + \sqrt{d} \cdot (m_n \cdot q_{n-1} + k_n \cdot q_{n-2}) = (m_n \cdot p_{n-1} + k_n \cdot p_{n-2}) + \sqrt{d} \cdot p_{n-1}$$

Equalizing both sides in the previous equation terms with no $\sqrt{d}$ and those who are next to $\sqrt{d}$:

$$\begin{cases} d \cdot q_{n-1} = m_n \cdot p_{n-1} + k_n \cdot p_{n-2} \\ p_{n-1} = m_n \cdot q_{n-1} + k_n \cdot q_{n-2} \end{cases}$$

Multiplying the first equation by q_{n-1}, the second one by p_{n-1} and subtracting the two results:

$$p_{n-1}^2 - d \cdot q_{n-1}^2 = k_n \cdot (p_{n-1} q_{n-2} + q_{n-1} p_{n-2})$$

Finally, we apply lemma 11.2 for $n-1$, $p_{n-1} q_{n-2} + q_{n-1} p_{n-2} = (-1)^{n-2}$ and we find the desired result $p_{n-1}^2 - d \cdot q_{n-1}^2 = (-1)^n \cdot k_n$ $\qquad\square$

n	a_n	p_n	q_n	$p_n^2 - 7 \cdot q_n^2$	k_n	r_n
0	2	2	1	$2^2 - 7 \cdot 1^2 = -3$	1	$(0 + \sqrt{7})/1$
1	1	3	1	$3^2 - 7 \cdot 1^2 = 2$	3	$(2 + \sqrt{7})/3$
2	1	5	2	$5^2 - 7 \cdot 2^2 = -3$	2	$(1 + \sqrt{7})/2$
3	1	8	3	$8^2 - 7 \cdot 3^2 = 1$	3	$(2 + \sqrt{7})/3$
4	4	37	14	$37^2 - 7 \cdot 14^2 = -3$	1	$(2 + \sqrt{7})/1$
5	1	45	17	$45^2 - 7 \cdot 17^2 = 2$	3	$(2 + \sqrt{7})/3$
6	1	82	31	$82^2 - 7 \cdot 31^2 = -3$	2	$(1 + \sqrt{7})/2$
7	1	127	48	$127^2 - 7 \cdot 48^2 = 1$	3	$(1 + \sqrt{7})/3$

$\cdots$

In the previous table we can see several of the results we have proved before, applied to the example $\alpha = \sqrt{7}$. First, in the column on the right we have the residuals, which, due to being bounded, must be repeated at some point, with r_5 being equal to r_1 as the first coincidence (as corollary 11.2 states, r_1 is involved in the first repetition), which causes the period to end there (and NOT when the first coefficient a_n is repeated, as someone can mistakenly think). Therefore, the period is equal to four and, as we saw in the same corollary, $k_4 = 1$ and $a_4 = 2a_0$.

The column on the left of the residuals has the values of k_n (that is, the denominators of the residuals), which, as we have just proved, must coincide with the values of $p_n^2 - 7 \cdot q_n^2$, although displaced one position and with a different sign alternately. Therefore, for $n = 3$ we already have the first solution of Pell's equation (since $k_4 = 1$ and have a positive sign): $8^2 - 7 \cdot 3^2 = 1$.

We just need to write and prove the generic case to end our solution.

COROLLARY 11.3. *Let $(a_0, \overline{a_1, ..., a_N})$ be the coefficients of the continued fraction of $\alpha = \sqrt{d}$. If N is even, then $p_{bN-1}^2 - d \cdot q_{bN-1}^2 = 1$ for any positive integer b; otherwise, if N is odd, then $p_{2bN-1}^2 - d \cdot q_{2bN-1}^2 = 1$ for any positive integer b.*

PROOF. By the previous theorem we have $p_{n-1}^2 - d \cdot q_{n-1}^2 = (-1)^n \cdot k_n$, but when n is multiple of the period, $n = bN$ (b any positive integer), we proved in corollary 11.2 that $k_{bN} = 1$ (we saw that $k_N = 1$, but the rest is inferred from it being a periodic continued fraction), which means that $p_{bN-1}^2 - d \cdot q_{bN-1}^2 = (-1)^{bN}$.

If N is even then $(-1)^{bN} = 1$ and the proof is finished; if N is odd, then $(-1)^{bN} = 1$ equals to 1 only when b is even, hence the need to use even values. $\square$

Note: For no other value of n we are going to achieve that $k_n = 1$, since when $r_n = m_n + \sqrt{d}$ (that is, when $k_n = 1$) we have that $a_n = m_n + a_0$ and:

$$r_{n-1} = \cfrac{1}{r_n - (m_n + a_0)} = \cfrac{1}{\sqrt{d} - a_0} = r_1$$

and that means that n must be a multiple of the period

For our example $\alpha = \sqrt{7}$ we had that its continued fraction is $(2, \overline{1, 1, 1, 4})$. That means that its period is 4, an even number, so $p_{4b-1}^2 - 7 \cdot q_{4b-1}^2 = 1$ for all positive integers b: we have, therefore, infinite solutions to Pell's equation $x^2 - 7y^2 = 1$ with convergent values of indices $\{3, 7, 11, ...\}$. The first solution is attained with the values of p_3, q_3 ($8^2 - 7 \cdot 3^2 = 1$).

For another example like $\sqrt{73}$ the numbers involved are larger, but the theory to apply is the same. First we must find its continued fraction, which is periodic and of type $(a_0, \overline{a_1, ..., a_N})$, as we saw in corollary 11.2. It is calculated as explained in its definition, and we get $(8, \overline{1, 1, 5, 5, 1, 1, 16})$. That means that its period is 7, an odd number, so $p_{14b-1}^2 - 73 \cdot q_{14b-1}^2 = 1$ for any positive integer b. Then we must look for the convergents (we also saw how to calculate them effectively with the equations (11.3)) and study the indices $\{13, 27, 41, ...\}$, which lead us to the solutions. The first solution is attained with the values of p_{13}, q_{13}, which turn out to be "large" numbers:

$$(2281249)^2 - 73 \cdot (267000)^2 = 1$$

FINAL REMARKS

- It can be verified that the solutions to Pell's equation calculated with continued fractions are, in fact, the **unique** solutions of the equation. We leave to the interested reader the task of learning about this statement.

- Once we have found the way to solve the equation $x^2 - d \cdot y^2 = 1$, automatically many other second degree Diophantine equations can be solved from it. Gauss, in his masterpiece "*Disquisitiones Arithmeticae*" (written at the age of 21) devotes an entire chapter of the book (hundreds of pages) to explain the complete method.

For example, with the explanations of Gauss it is possible to find the integer solutions of the equation $x^2 + 8xy + y^2 + 2x - 4y + 1 = 0$, applying convenient changes of variables that transform it into Pell's equation $t^2 - 15 \cdot u^2 = 1$, which allow us to find the solutions:

$$(x, y) = [(1, -2), (-1, 0), (-1, 12), (13, -98), (13, -2), (-97, 12), ...]$$

Chapter 12

Circular billiard

(Alhazen – 1015)

PROBLEM

To find how to hit a billiard ball in a circular table in order to hit another one after the bounce.

HISTORY

This problem was proposed and solved by the Arab mathematician Abu Ali al Hassan ibn al Hassan ibn Alhaitham (965 – 1039), whose name was transformed into Alhazen by the translators of his work "Optics". In the book the problem is stated in the following (equivalent) way: "To find the point in a spherical mirror where a ray of light that comes from a given point is reflected at another given point".

Figure of Alhazen
10 dinars note (Iraq)

Alhazen wrote the first treatise on lenses, studied the reflection and refraction of light, understood the formation of the rainbow and defended the idea of the finite thickness of the atmosphere, among other achievements.

A complete series of famous mathematicians took this problem after Alhazen, among them Huygens, Barrow, L'Hôpital, Riccati and Quételet.

SOLUTION

Let C be the circumference that forms the band of the billiard table, M the center of the circle, r its radius, P_1 and P_2 the original positions of the two balls (we consider them as points without thickness). Let us take M as the origin of a system of perpendicular coordinates $x - y$. Without loss of generality, we can assume that $r = 1$, that P_1 is in the first quadrant of the coordinate system and that its distance to M is greater (or equal) than the distance of P_2 to M (otherwise, we would take P_2 as P_1). We can also assume that P_2 is in the fourth quadrant and that the axis x is the bisector of the angle $\widehat{P_1 M P_2}$. (The reader should check that any original position

69

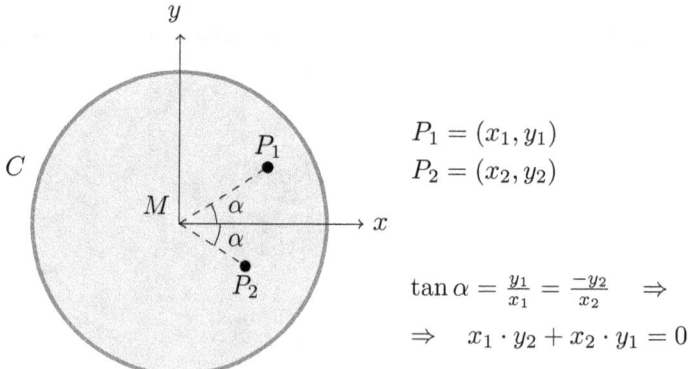

$$P_1 = (x_1, y_1)$$
$$P_2 = (x_2, y_2)$$

$$\tan \alpha = \frac{y_1}{x_1} = \frac{-y_2}{x_2} \quad \Rightarrow$$

$$\Rightarrow \quad x_1 \cdot y_2 + x_2 \cdot y_1 = 0$$

Figure 12.1

of the balls can be converted, by means of a suitable rotation and a possible symmetry, into this proposed case).

Under these conditions, the coordinates of $P_1 = (x_1, y_1)$ and $P_2 = (x_2, y_2)$ fulfill the following equations:

(12.1)
$$\begin{cases} 0 \le x_2 \le x_1 \le r \\ 0 \le -y_2 \le y_1 \le r \end{cases}$$

(12.2)
$$x_1 \cdot y_2 + x_2 \cdot y_1 = 0$$

The equation (12.2) follows from the fact that the x axis is the bisector of the angle $\widehat{P_1 M P_2}$ (see figure 12.1).

Suppose now that the point sought in the billiard band is the point O with coordinates (x, y). In that case, the property that the ball will bounce in the band at point O as it would in a billiard table is equivalent to say that the angle $\widehat{P_1 O M}$ is equal to the angle $\widehat{P_2 O M}$ (that is, the angle of incidence is equal to the angle of reflection). Let ϕ be the value of this angle.

In addition, we denote the angles of the segments $\overline{P_1 O}$, $\overline{P_2 O}$ and $\overline{MO}$ with the x axis as α_1, α_2 and β, respectively.

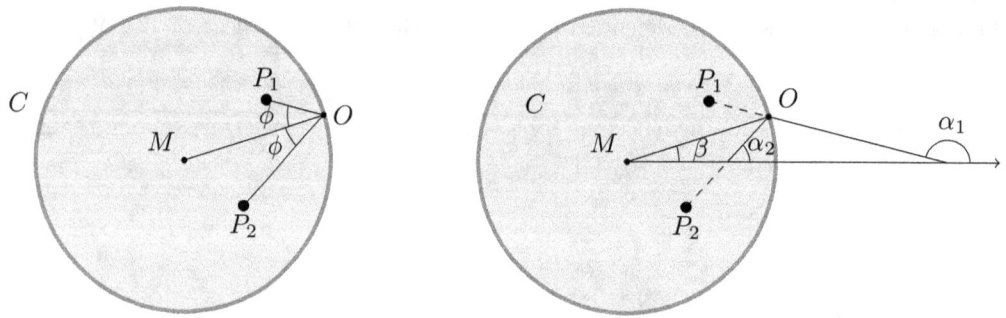

Figure 12.2

As we can see in figure 12.2, from the triangle of vertices M, O and the point of intersection of $\overline{P_1 O}$ with x axis we deduce that:

$$(180 - \alpha_1) + \beta + (180 - \phi) = 180 \quad \Rightarrow \quad \phi = (180 - \alpha_1) + \beta \quad \Rightarrow$$

$$(12.3) \qquad \Rightarrow \qquad \tan\phi = \frac{\tan(180 - \alpha_1) + \tan\beta}{1 - \tan(180 - \alpha_1) \cdot \tan\beta} \qquad \Rightarrow \qquad \tan\phi = \frac{-\tan\alpha_1 + \tan\beta}{1 + \tan\alpha_1 \cdot \tan\beta}$$

In a similar fashion, from the triangle of vertices M, O and the point of intersection of $\overline{P_2O}$ with the x axis we deduce that:

$$(180 - \alpha_2) + \beta + \phi = 180 \qquad \Rightarrow \qquad \phi = \alpha_2 - \beta \qquad \Rightarrow$$

$$(12.4) \qquad \Rightarrow \qquad \tan\phi = \frac{\tan\alpha_2 - \tan\beta}{1 + \tan\alpha_2 \cdot \tan\beta}$$

Finally, considering that (see figure 12.3):

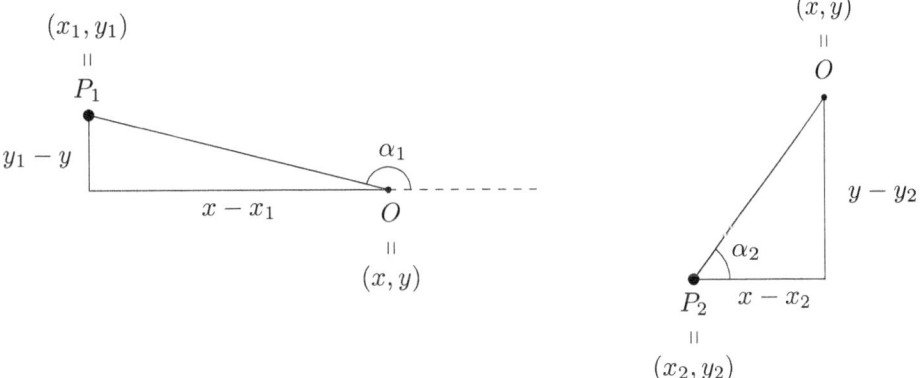

Figure 12.3

$$\tan\alpha_1 = \frac{y - y_1}{x - x_1} \qquad \tan\beta = \frac{y}{x} \qquad \tan\alpha_2 = \frac{y - y_2}{x - x_2}$$

we can equalize equations (12.3) and (12.4), substituting in them the values found previously:

$$\frac{-\frac{y-y_1}{x-x_1} + \frac{y}{x}}{1 + \frac{y-y_1}{x-x_1} \cdot \frac{y}{x}} = \frac{\frac{y-y_2}{x-x_2} - \frac{y}{x}}{1 + \frac{y-y_2}{x-x_2} \cdot \frac{y}{x}} \qquad \Rightarrow$$

$$(12.5) \qquad \Rightarrow \qquad \frac{x \cdot y_1 - y \cdot x_1}{x^2 + y^2 - x \cdot x_1 - y \cdot y_1} = \frac{-x \cdot y_2 + y \cdot x_2}{x^2 + y^2 - x \cdot x_2 - y \cdot y_2}$$

This equation (12.5) has been inferred assuming that the solution point O is in the first quadrant, although it would be reached too by assuming to be in any other quadrant. Operating with equation (12.5) we get to:

$$(12.6) \qquad H \cdot (x^2 - y^2) - 2Kxy + (x^2 + y^2) \cdot [hy - kx] = 0$$

where $H = x_1 \cdot y_2 + x_2 \cdot y_1$, $K = x_1 \cdot x_2 - y_1 \cdot y_2$, $h = x_1 + x_2$ and $k = y_1 + y_2$

However, we have seen in (12.2) that $H = 0$, so finally we can say that the solution points O to our problem meet the equation:

(12.7)
$$-2Kxy + (x^2 + y^2) \cdot [hy - kx] = 0$$

So far we have not applied that O must be a point of the circumference C, so the solution points are those that satisfy the following system of equations:

$$\begin{cases} -2Kxy + (x^2 + y^2) \cdot [hy - kx] = 0 \\ x^2 + y^2 = 1 \end{cases} \quad \Rightarrow \quad \begin{cases} -2Kxy + [hy - kx] = 0 \\ x^2 + y^2 = 1 \end{cases}$$

That is, the points sought correspond to the intersection between the quadratic equation:

(12.8)
$$-2Kxy + [hy - kx] = 0$$

and the unit circumference that corresponds to the band of the pool table. Therefore, we must first study what kind of conics corresponds to equation (12.8).

First we must take into account, for the properties defined in (12.1), that the previously defined variables K, h, k are all non-negative (for K and h the deduction is trivial, while k is positive by the imposition that P_1 is at greater distance than P_2 from the origin and by the fact that the X axis is the angle bisector of $\widehat{P_1 M P_2}$). Next, we write the equation (12.8) in its matrix form, as it is known by conics theory:

(12.9)
$$(x \quad y \quad 1) \cdot \begin{pmatrix} 0 & -K & -k/2 \\ -K & 0 & h/2 \\ -k/2 & h/2 & 0 \end{pmatrix} \cdot \begin{pmatrix} x \\ y \\ 1 \end{pmatrix} = 0$$

Let D_3 be the value of the determinant of the 3x3 matrix in (12.9), D_2 the 2x2 determinant of the first 2 rows and columns and D_1 the sum of the values in the diagonal of the matrix. Then:

$$D_3 = K \cdot \frac{kh}{2} \qquad D_2 = -K^2 \qquad D_1 = 0$$

General case

In the general case ($K \neq 0, k \neq 0, h \neq 0$), that is, the billiard balls are neither in the center nor in the same diameter, we have that $D_3 > 0$, $D_2 < 0$ and $D_1 = 0$, which in conic theory defines an **equilateral hyperbola**.

In fact, assuming this general case, applying to equation (12.8) the following change of coordinates:

(12.10)
$$\begin{cases} x' = x - \frac{h}{2K} \\ y' = y + \frac{k}{2K} \end{cases}$$

we have that equation (12.8) is converted into another one with the form of an equilateral hyperbola:

(12.11)
$$x' \cdot y' = -\frac{k \cdot h}{4K^2}$$

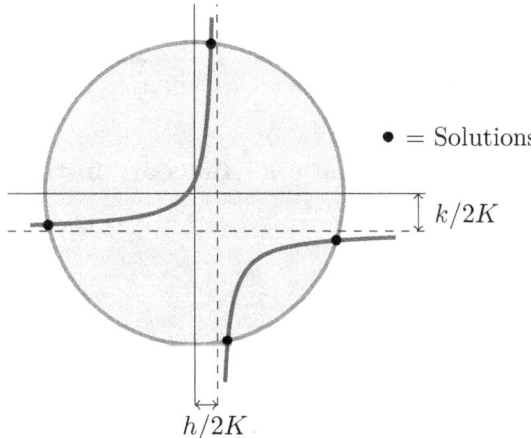

Figure 12.4

Therefore, the general solution to the problem (when none of the balls are in the center of the table and they are not in the same diameter) is the following: "**Solutions where we must point the ball to are the intersections of the hyperbola (12.11) with the circumference that defines the table**".

FINAL REMARKS

We can now study some particular cases: the first three cases are not part of the general solution (because a ball is in the center of the table or the balls are in the same diameter), and we will complete the solutions of the problem; the last case is a particular case of the general solution that is especially interesting (both balls are at the same distance from the center of the table).

Case not covered in the general solution: (Balls are in the same diameter with the center of the table between them)

In this case, $x_1 = x_2 = 0$, which implies that $h = 0$ and $y_1 \neq 0 \neq y_2$, i.e., we have both balls on the y axis, **each one on a different side with respect the center of the table** but none in the center (Figure 12.5).

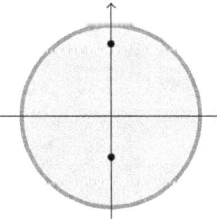

Figure 12.5

In this case, we have that equation (12.8) becomes $-2Kxy - kx = 0$, which has points in the lines $x = 0$ and $y = -k/(2K)$ as solutions. The intersection with the unit circle gives us the two obvious points $(0, 1)$, $(0, -1)$ and two more points in the line $y = -k/(2K)$, that is, in the semicircle of negative ordinates, where the ball is closest to the center.

Keep in mind that, in some cases, this line has **NO** intersection with the unit circle (for example, $y_1 = 1/4$ and $y_2 = -3/8$ give as a solution in the line $y = -2$), while the intersection exists for other cases (for example, $y_1 = 1/2$ and $y_2 = -3/8$ yield a solution in the line $y = -1/3$).

Case not covered in the general solution: (Balls are in the same diameter with the center of the table not between them)

In this case, $y_1 = y_2 = 0$, which implies that $k = 0$ and $x_1 \neq 0 \neq x_2$, i.e, we have both balls in the x axis, **both in the same side of the table**, none of them in the center of the table (Figure 12.6).

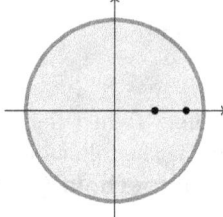

Figure 12.6

In this case, we have that equation (12.8) becomes $-2Kxy + hy = 0$, which has points of the lines $y = 0$ and $x = h/2K$ as solutions. The intersection with the unit circle yields the two obvious points $(1, 0)$, $(-1, 0)$ and two more points on the line $x = h/2K$.

But unlike the previous case this line **NEVER** has an intersection with the unit circumference, because applying the properties of (12.2) and the property that the arithmetic mean is greater than the geometric mean for positive numbers it holds that:

$$\frac{h}{2K} = \left(\frac{x_1 + x_2}{2} \right) \cdot \frac{1}{x_1 \cdot x_2} > \sqrt{x_1 \cdot x_2} \cdot \frac{1}{x_1 \cdot x_2} = \frac{1}{\sqrt{x_1 \cdot x_2}} > 1$$

That is, we will never have other solutions different from the two trivial ones.

Case not covered in the general solution: (A ball is in the center of the table)

In this case, $x_2 = y_2 = 0$, which implies that $K = 0$ and $x_1 \neq 0 \neq y_1$, i.e., we have one ball in the center of the table (figure 12.7).

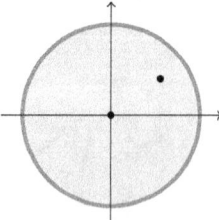

Figure 12.7

In this case, we have that equation (12.8) becomes $hy - kx = 0$, which is precisely the line that joins both balls. Therefore, there are only two obvious solutions: to hit a ball following the line that joins both (theoretically, in both directions).

74

Particular case of the general solution: (Balls are at the same distance from the center)

Finally, suppose the interesting case in which the distance of the balls to the center of the table is identical, but within the general case. That is, $x_1 = x_2 \neq 0$, $y_1 = -y_2 \neq 0$ and, if we define c as the distance of one of the balls to the center of the table, we have $c^2 = x_1^2 + y_1^2 = x_2^2 + y_2^2 > 0$.

This case falls within the general case, so the solutions (places where we must point the ball to) are those indicated in Figure 12.3. What we are going to see here is an additional curious property that is fulfilled in this case, and that I have considered convenient to explain and prove.

In this case, we have that the equation (12.8) becomes $-2c^2xy + 2x_1y = 0$, with two obvious solutions in the line $y = 0$ and two solutions in the line $x = x_1/c^2$. Depending on the positions of the balls this line will have or not intersection with the unit circumference: if there is no intersection, only the two trivial solutions exist, so let us suppose now that $x_1/c^2 < 1$, that is, $c^2 > x_1$.

Now let us look for the circumference that has center in the axis of the abscissas and that pass through M and through the two previous solutions (intersections of the line $x = x_1/c^2$ with the unit circumference, which we name Q points). Suppose that the center of this circumference is at point $(r_1, 0)$ and let N be the opposite point of M on the circumference (see figure 12.8).

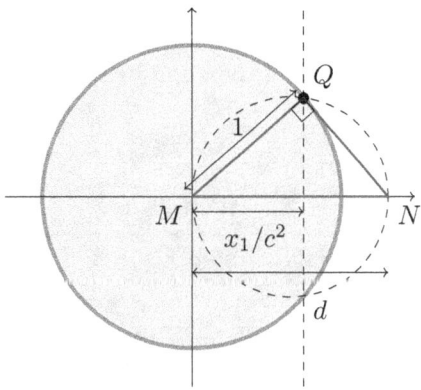

Figure 12.8

Since Q belongs to the unit circle we have $\overline{MQ} = 1$ and its projection to the axis of the abscissas is at distance x_1/c^2 from M, as we have seen previously. But in addition, $\widehat{MQN}$ is a right angle (because Q is a point of the circumference with diameter $\overline{MN}$). Then we can apply the right angle altitude theorem to MQN to deduce:

$$\overline{MN} = \frac{\overline{MQ}^2}{(x_1/c^2)} = \frac{1}{(x_1/c^2)} = \frac{c^2}{x_1}$$

Therefore, we have that the circumference of center $(c^2/2x_1, 0)$ and radius $c^2/(2x_1)$ passes through both M and Q. But in addition, we will prove that it also passes through the points P_1 and P_2, where the balls are. For this, we must show that the distance d from $(c^2/2x_1, 0)$ to (x_1, y_1) is equal to the radius, that is, $c^2/(2x_1)$. But this is easily seen:

$$d^2 = \left(\frac{c^2}{2x_1} - x_1\right)^2 + y_1^2 = \left(\frac{c^2}{2x_1}\right)^2 - c^2 + x_1^2 + y_1^2 = \left(\frac{c^2}{2x_1}\right)^2$$

so we have proved that M, P_1 and Q are in the same circle.

So the four vertices of the quadrilateral MP_1QP_2 lie in the same circle, and a quadrilateral satisfying that property is known as a **cyclic quadrilateral** (see figure 12.9). Ptolemy's theorem says that, in a cyclic quadrilateral, the sum of the two products of lengths of opposite sides is equal to the product of the lengths of the diagonals. Therefore, in this case:

$$\overline{P_1Q} \cdot \overline{P_2M} + \overline{P_2Q} \cdot \overline{P_1M} = \overline{MQ} \cdot \overline{P_1P_2} \qquad \Rightarrow$$

(12.12)
$$\Rightarrow \qquad (\overline{P_1Q} + \overline{P_2Q}) \cdot c = 2y_1$$

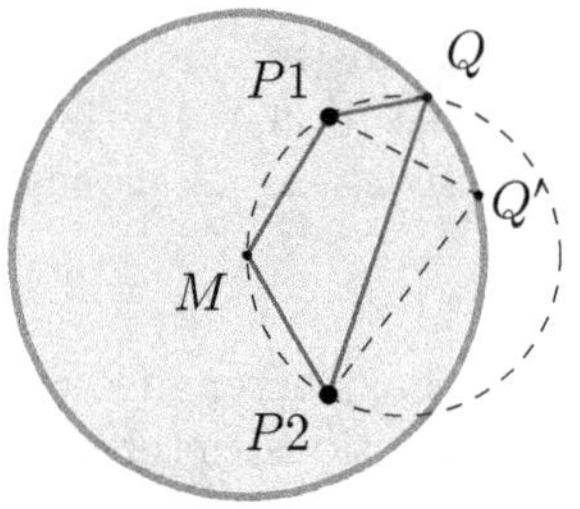

Figure 12.9

For another point Q' of C we have that the quadrilateral $MP_1Q'P_2$ is not cyclic and, also by Ptolemy's Theorem, the sum of the product of the opposite sides is **greater** than the product of its diagonals. So:

(12.13)
$$\Rightarrow \qquad (\overline{P_1Q'} + \overline{P_2Q'}) \cdot c > 2y_1$$

From (12.12) and (12.13) it follows that $\overline{P_1Q} + \overline{P_2Q} < \overline{P_1Q'} + \overline{P_2Q'}$ and hence the curious property that, "**in the case that the balls are the same distance from the center of the table, the non-trivial solutions of the Alhazen problem (if they exist) are those that fulfill the property that the sum of distances to both balls is minimal**". This property (minimal path) is identical to the minimum sum of distances in a typical billiard table.

Chapter 13

Binomial expansion

(Khayyam – 1090)

PROBLEM

To obtain the coefficients of the n-th power of the binomial $a + b$ in terms of powers of a and b, when n is any positive integer number.

HISTORY

The binomial theorem (or binomial expansion) was probably discovered by the Persian Omar Khayyam (1048 – 1131), considered one of the best astronomers and mathematicians of medieval times. At least he prided himself on having discovered the expansion "for every exponent (positive integer) n, which nobody had done before".

Statue of Omar Khayyam
Illustrious Persians pavilion (United Nations, Wien)
Photo: Yamaha5 (Wikimedia Commons)

This theorem coexists with any student of sciences during most of his studies, but perhaps many of them have never been able to see its proof.

SOLUTION

$$(a + b)^n = (a + b)(a + b) \cdots (a + b)$$

As it is well-known, the previous product is equivalent to a sum of terms of type $Ca^r b^s$, where C is a constant to be determined that depends on n, r and s, where $r + s = n$. The problem is to determine the so-called *polynomial coefficient* C, that is, to answer the question: How many times does the product $a^r b^s$ appear in the binomial expansion?

Let us first see an example. Suppose we want to know how many times the term $a^4 b^2$ appears in the expansion of $(a + b)^6$, that is $n = 6$, $r = 4$ and $s = 2$. The formula that we have to find has to yield, in this case, a total of $C = 15$ cases, which are those that are achieved with these combinations:

$$
\begin{array}{ccccc}
aaaabb & aaabab & aabaab & abaaab & baaaab \\
aaabba & aababa & abaaba & baaaba & aabbaa \\
ababaa & baabaa & abbaaa & babaaa & bbaaaa
\end{array}
$$

where in each of them the order of the factor that was chosen for each parenthesis has been conserved (that is, for example, the term *abaaba* comes from choosing $(\underline{a} + b)(a + \underline{b})(\underline{a} + b)(\underline{a} + b)(a + \underline{b})(\underline{a} + b)$).

To find the formula for the calculation of C, let us think in the following way: suppose we have 4 books of different colors with the same title "a", and 2 books of different colors with the same title "b". Let us think of all possible ways to place them on a shelf where there are exactly 6 spots: in the first spot we can place any of the 6 books (6 possibilities); once placed the first, now we have 5 possibilities (among those that remain) to put the following one; for the third we have 4 possibilities; and so on until only one possibility remains (the last book) to place in the last spot.

That is, there are $6 \cdot 5 \cdot 4 \cdot 3 \cdot 2 \cdot 1 = 720$ possibilities to put the 6 books in the 6 spots of the shelf. This multiplication of a positive integer number n by all its smaller positive integer numbers was abbreviated, centuries ago, with the notation $n!$ and it was called the **factorial of** n, so in our example there is a total of 6! book possibilities.

An example of a **book layout**, with its colors, out of the possible $6! = 720$:

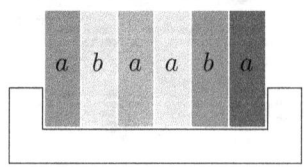

Figure 13.1

The above example can be associated with the **factor order** *abaaba*. The problem is that there are many book dispositions that, when removing colors, yield the same order of factors. For example, the order of books:

produces the same order of factors *abaaba*. But at least we know that for each order of factors there are a number of identical book dispositions, so now we have to take an order of factors,

78

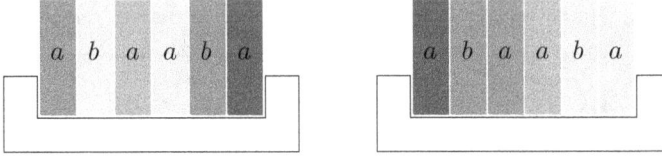

Figure 13.2

such as *abaaba* and deduce how many book dispositions are related. If there were, for example, 10 provisions for each, that means there is a total of 720/10 different ways of ordering the factors.

Therefore, we have to deduce how many book dispositions lead to the same order of factors. Suppose we forget for a moment that the "b" title books have a color. In that case, all the possibilities to change places between all the books of title "a" produce the same disposition of books: but the books of title "a" are 4 and there are 4 spots where to place them (you have to change among them), so, as we have seen before, that's 4! = 24 different possibilities.

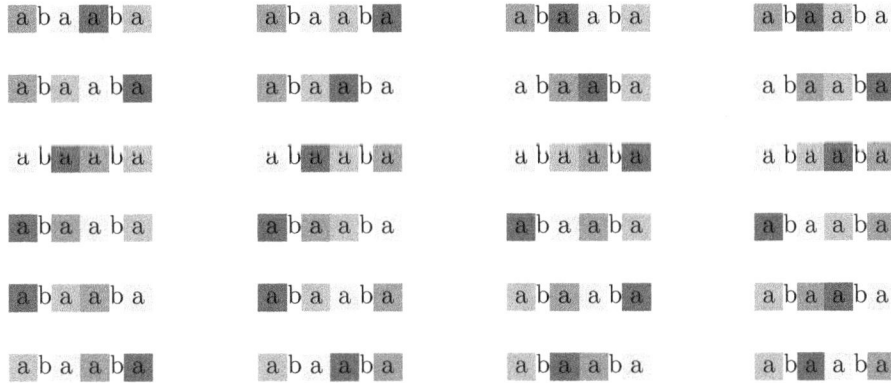

Therefore, if we divide 720/24 =30 we are removing the colors of "a" and all the orders of factors in which "b" are still colored remain.

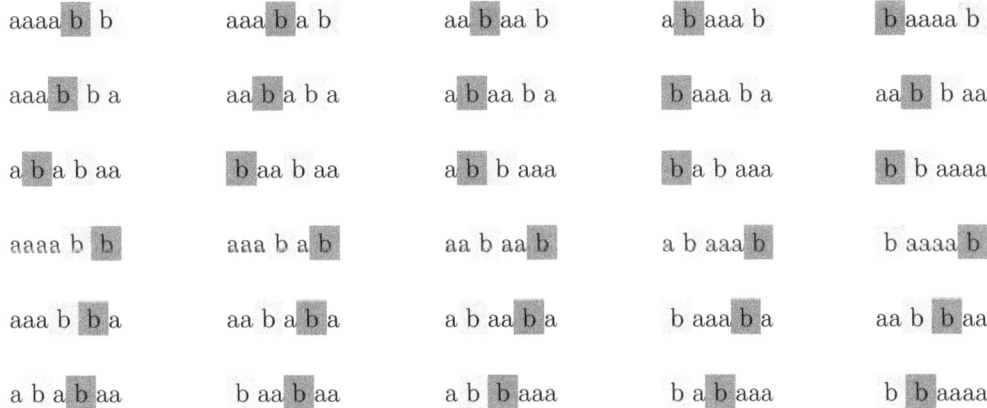

How many of them lead to the same order of factors when we remove the color of the "b" books? For the same reason as before, 2! = 2 possibilities (there are two books for two positions). Therefore, there is a total of 30/2 = 15 orders of possible factors, which is the number that we had calculated at the beginning (simply writing all the cases).

The reasoning is identical for the general case: first there are a total of $n!$ book dispositions, then we divide by $r!$ to remove the colors from the "a" books and then divide by $s!$ to remove the colors from the "b" books. The reasoning is curious, because first we calculate many more possibilities (when adding colors) than what was requested in the problem, and then reduce them (by removing some of them); however, in my opinion this is the simplest way to deduce the formula.

Therefore, the formula for C is:

$$C = \frac{n!}{r! \cdot s!} = \frac{n!}{r! \cdot (n-r)!}$$

where $s = n - r$. The previous expression has a name and a symbol, due to its importance in probability theory. The name is "**binomial of n over r**" (precisely because it came from the exponentiation of a binomial of factors) and the symbol is:

$$\binom{n}{r}$$

If we are careful to define $0! = 1$ (it is just a convention to avoid later problems in the formulas), we finally recover the famous way of writing the binomial expansion:

$$(a + b)^n = \sum_{0 \le r \le n} \binom{n}{r} a^r b^{n-r} = \binom{n}{0} a^n + \binom{n}{1} \cdot a^{n-1}b + \cdots + \binom{n}{n-1} \cdot ab^{n-1} + \binom{n}{n} b^n$$

In our example:

$$(a + b)^6 = \binom{6}{0} a^6 + \binom{6}{1} a^5 b^1 + \binom{6}{2} a^4 b^2 + \binom{6}{3} a^3 b^3 + \binom{6}{4} a^2 b^4 + \binom{6}{5} a^1 b^5 + \binom{6}{6} b^6 =$$

$$= \frac{6!}{6! \cdot 0!} a^6 + \frac{6!}{5! \cdot 1!} a^5 b^1 + \frac{6!}{4! \cdot 2!} a^4 b^2 + \frac{6!}{3! \cdot 3!} a^3 b^3 + \frac{6!}{2! \cdot 4!} a^2 b^4 + \frac{6!}{1! \cdot 5!} a^1 b^5 + \frac{6!}{0! \cdot 6!} b^6 =$$

$$= a^6 + 6a^5 b + 15a^4 b^2 + 20a^3 b^3 + 15a^2 b^4 + 6ab^5 + b^6$$

FINAL REMARKS

– The formula can be easily inferred for the n-th power of the expression $(a + b + c + \cdots)$. With three variables, for example, we have:

$$(a + b + c)^n = \sum_{r+s+t=n} \frac{n!}{r!s!t!} a^r b^s c^t$$

where the sum includes all possible terms with r, s, t as non-negative integers that fulfill the relation $r + s + t = n$.

– Binomials $\binom{n}{r}$ are the famous coefficients in Pascal's triangle, built with the well-known formula:

$$\binom{n}{r} + \binom{n}{r+1} = \binom{n+1}{r+1}$$

which can be easily deduced with what we have seen before:

$$\binom{n}{r} + \binom{n}{r+1} = \frac{n!}{r! \cdot (n-r)!} + \frac{n!}{(r+1)! \cdot (n-r-1)!} =$$

$$= \frac{(r+1) \cdot n!}{(r+1)! \cdot (n-r)!} + \frac{n! \cdot (n-r)}{(r+1)! \cdot (n-r)!} = \frac{[(r+1) + (n-r)] \cdot n!}{(r+1)! \cdot (n-r)!} =$$

$$= \frac{(n+1) \cdot n!}{(r+1)! \cdot (n-r)!} = \frac{[(n+1)!}{(r+1)! \cdot (n-r)!} = \binom{n+1}{r+1}$$

That is, in Pascal's triangle (see figure 13.3), any term (except the ends of each row, which are always equal to 1) is equal to the sum of the two closest values of the previous row. This allows to find the coefficients of each row based on the previous row, which is interesting but not very effective if we want to calculate a coefficient of a row that is far away from the vertex of the triangle.

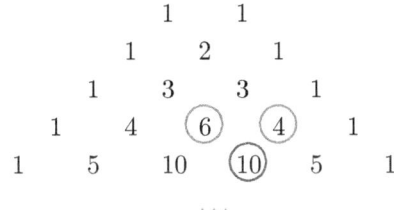

Figure 13.3

Chapter 14

Maximum vision of the rings of Saturn

(Muller – 1471)

PROBLEM

To determine the latitude from where an observer on the surface of Saturn can see its rings with the maximum possible angle.

HISTORY

After the impressive development of Mathematics in antiquity (by Egyptians, Chinese and Greeks, especially), the advances during the following centuries were of lesser importance.

Statue dedicated to Johannes Muller
Konigsberg (Baviera)
Photo: Tilman2007 (Wikimedia Commons)

In fact, it is curious to note that the first problem of finding extreme points (after antiquity) did not appear until 1471, when the mathematician Johannes Muller (also known by his Latin nickname "Regiomontanus") proposed to Professor Christian Roder the problem that concerns us (expressed in other conditions but equivalent to the rings of Saturn, a more striking statement).

The solution presented here, of great simplicity and elegance, is not Muller's original, but one found in volume XXIII of the book "Zeitschrift für Mathematik und Physik".

SOLUTION

Saturn can be considered a sphere with a radius equal to 56900 kilometers and its rings can be considered as one, circular, in the plane of Saturn's equator, with an inner and an outer radius of 88500 and 138800 kilometers. Obviously, the problem can be simplified, by symmetry, taking any meridian of Saturn and considering only one hemisphere, so the problem to be studied is shown in Figure 14.1.

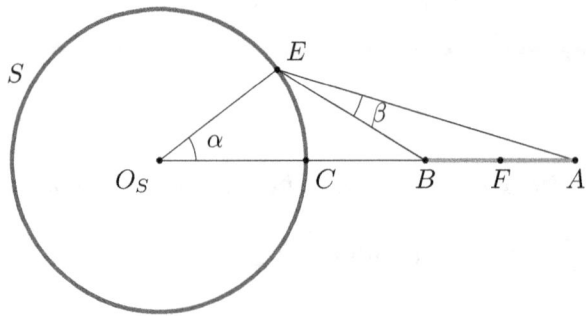

Figure 14.1

Let S be any meridian of Saturn, O_S be the center of the planet and r_S be its radius ($r = 56900$ km). Let A (resp., B) be the outer (resp., inner) point of the ring that is in the plane containing S ($\overline{O_S A} = 138800$ km, $\overline{O_S B} = 88500$ km). The line that passes through A and B also passes through O_S (since we have assumed that the rings are in the plane of the equator), and let C be the point where this line crosses the meridian S between A and O_S. Finally, let F be the midpoint of the segment $\overline{AB}$, E be any point of the meridian S, α be the angle $\widehat{EO_S C}$, that is, the solution to the problem, and β be the angle $\widehat{BEA}$, which we want to maximize.

PROPOSITION 14.1. *With the notation used, the point E_0 from where we observe the segment $\overline{AB}$ with the greatest angle would be the point of tangency between S and the circumference that passes through A, B and that is tangent to S.*

PROOF. Let T be a circumference of radius r_T that passes through A and B (its center O_T must be in the mediatrix of $\overline{AB}$ that passes through F), and let G be a point of T that is in the half-plane defined by the line that passes through $\overline{AB}$ and contains O_T. Using the property of the central angle (see theorem 25.1), the value of the angle $\widehat{BGA}$ is half the value of the angle $\widehat{BO_T A}$, as we can see in figure 14.2.

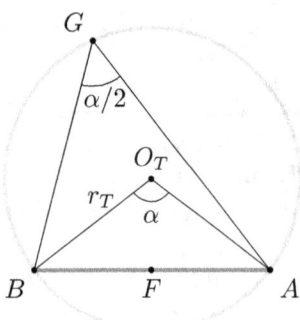

Figure 14.2

In addition, there is a trigonometric relationship between the angle $\widehat{BO_T A}$ and the radius r_T, since $\sin(\widehat{BO_T A}/2) = \overline{AF}/r_T$, i.e., $\widehat{BO_T A}/2 = \arcsin(\overline{AF}/r_T)$. Using this observation together with what we saw in the previous paragraph, we have that $\widehat{BGA} = \arcsin(\overline{AF}/r_T)$ and it is deduced, therefore, that as r_T increases the angle $\widehat{BGA}$ decreases.

Imagine now all the possibilities for the circumference T:

- If T does not intersect S, there is nothing to study.
- If T is tangent to S, let G' be the point of tangency between S and T (second part of figure 14.2); the angle to study is $\widehat{BG'A}$.

- If T cuts S at two points called G_1 and G_2 (see figure 14.3), the angle $\widehat{BG_1A}$ (or $\widehat{BG_2A}$) is smaller than the angle $\widehat{BG'A}$, since the radius of the circumference T that cuts S in G_1 and G_2 is greater than the radius of the circumference T tangent to S. Since the radius is greater, that means that the angle $\widehat{BG_1A}$ is smaller than the angle $\widehat{BG'A}$.

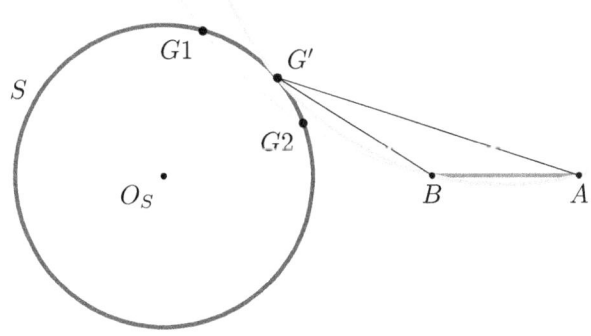

Figure 14.3

In conclusion, the angle $\widehat{BG'A}$ is the largest of those produced by the circumferences T that cut S, so it is our solution. $\qquad\square$

Once the solution is found, our last mission will be to write the formulas to define it accurately. However, before that it is necessary to introduce a new concept: the "power of a point" with respect to a circle.

DEFINITION 14.1. *Let S be a circumference of radius r and let P be a point outside it. Suppose a line that passes through P and intersects S in two points S_1 and S_2. We define the **power of P with respect to the circumference** S (and we write $P(S)$) as the value:*

$$P(S) = \overline{PS_1} \cdot \overline{PS_2}$$

PROPOSITION 14.2. *The power of P with respect to S does not depend on the chosen line. That is, if we chose any other line that passes through P and cuts S in two other points S_1' and S_2', then:*

$$\overline{PS_1} \cdot \overline{PS_2} = \overline{PS_1'} \cdot \overline{PS_2'}$$

PROOF. We are going to use theorem 2.2 (Stewart) in figure 14.4, where we have drawn the line (which we call "**central**") that passes through P and through the center O of the circumference (and cuts S at points S_1 and S_2), and another line that cuts S (at points S_1' and S_2'). If we show that the value of the power of P with respect to S is the same for both lines, then we will have shown that it will be the same for any other (it will always be equal to the value calculated with the "central" line).

Let d be the distance between P and S_1. Then, for the center line, the power of P is calculated as $\overline{PS_1} \cdot \overline{PS_2} = d \cdot (d + 2r)$. Therefore, we have to see that the power of P calculated with the other line ($\overline{PS_1'} \cdot \overline{PS_2'}$) yields the same value.

As we have said, we will use Stewart's theorem, taking the triangle formed by vertices P, O and S_2', and the segment $\overline{OS_1'}$. Let t be the distance between P and S_1', and let s be the distance between P and S_2'. Applying the theorem:

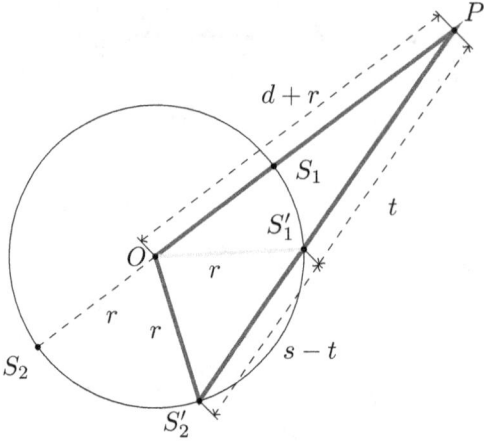

Figure 14.4

$$t \cdot r^2 + (s - t) \cdot (r + d)^2 = s \cdot (r^2 + (s - t) \cdot t) \quad \Rightarrow \quad (s - t) \cdot (r + d)^2 = (s - t) \cdot [r^2 + s \cdot t] \quad \Rightarrow$$

$$\Rightarrow \quad (r + d)^2 = r^2 + s \cdot t \quad \Rightarrow \quad s \cdot t = (r + d)^2 - r^2 = 2dr + d^2 = d \cdot (d + 2r)$$

The value of the power of P with respect to S using any line is equal to $\overline{PS_1'} \cdot \overline{PS_2'} = t \cdot s = d \cdot (d + 2r)$, the same calculated with the central line. □

COROLLARY 14.1. *Let $a = \overline{O_S A}$ and $b = \overline{O_S B}$. With the notation used so far, at the point E_0 where $\overline{AB}$ is observed with the greatest angle, the following equations are true:*

$$\cos \alpha = \frac{(a + b) \cdot r_S}{ab + r_S^2} \qquad and \qquad \sin \beta = \frac{(a - b) \cdot r_S}{ab - r_S^2}$$

PROOF. The solution found in the previous proposition indicates that E_0 is the point of tangency between S and the circumference T that passes through A and B and that is tangent to S. In this case we can perform the following calculations.

Taking the right triangles $O_S O_T F$ and $A O_T F$ we can deduce (see figure 14.5):

$$(14.1) \qquad \cos \alpha = \frac{\overline{O_S F}}{\overline{O_S O_T}} = \frac{(a + b)/2}{r_S + r_T} \qquad and \qquad \sin \beta = \frac{\overline{AF}}{\overline{AO_T}} = \frac{(a - b)/2}{r_T}$$

where we have applied the central angle theorem again to deduce that $\beta = \widehat{BE_0A} = \widehat{FO_TA}$.

In equations (14.1) all values are known except r_T. To find an equation that relates r_T to the rest of the variables, we will apply proposition 14.2. Specifically, we are going to apply that the power of the point O_S with respect to the circumference T is the same regardless of the line we choose from those that cut T at two points.

Specifically, we consider the line $\overline{O_S AB}$(which cuts T at points A and B) and the line $\overline{O_S E_0 Z}$ (which cuts T at points E_0 and Z - the opposite of E_0 with respect to the center of the circle). Applying proposition 14.2 for these two lines, we have:

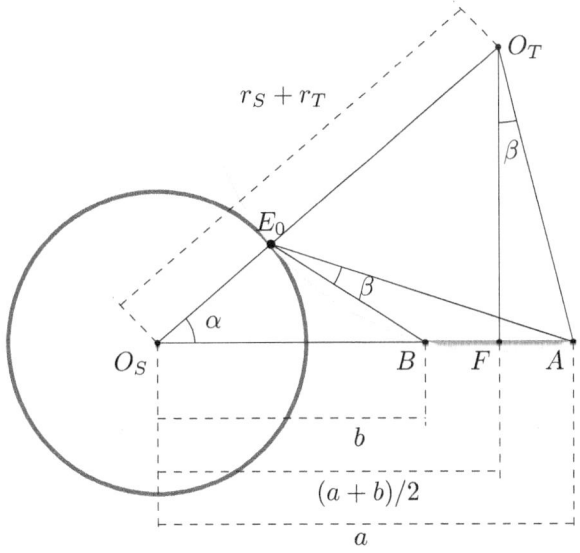

Figure 14.5

$$\overline{O_S A} \cdot \overline{O_S B} = \left(\overline{O_S O_T} - r_T\right) \cdot \left(\overline{O_S O_T} + r_T\right) \qquad \Rightarrow$$

$$(14.2) \qquad \Rightarrow \qquad a \cdot b = r_S \cdot (r_S + 2r_T) = r_S^2 + 2r_S r_T \qquad \Rightarrow \qquad r_T = \frac{ab - r_S^2}{2r_S}$$

Now it is only necessary to substitute the value of r_T found in (14.2) in each of the equations of (14.1) to find the desired relationships. $\qquad \square$

If we now apply the distances mentioned at the beginning (Saturn radius and the inner and outer radii of the rings), we see that the approximate result would be:

$$\cos\alpha = \frac{(138800 + 88500) \cdot 56900}{138800 \cdot 88500 + 56900^2} \approx 0.833$$

$$\sin\beta = \frac{(138800 - 88500) \cdot 56900}{138800 \cdot 88500 - 56900^2} \approx 0.316$$

$\alpha \approx 33.5°$ (latitude solution) $\qquad\qquad \beta \approx 18.4°$ (greatest angle of vision)

FINAL REMARKS

If Earth had a ring of thickness and distance in identical proportion to the ring of Saturn (proportional to the radii of both planets), the latitude solution and maximum angle of vision would have the same value. For example, the inhabitants of Tripoli (Libya) or Sydney (Australia), both at a latitude close to 33.5° (North in the case of Tripoli, South in the case of Sydney), would be fortunate to have that better view, while near the Equator (Singapore or Quito, for example) they would hardly see the edge of the ring.

Chapter 15

Solution of the cubic

(Tartaglia – 1530)

PROBLEM

To find a general procedure to solve third-degree equations with complex coefficients $x^3 + px^2 + qx + r = 0$.

HISTORY

The resolution of second-degree equations with real coefficients has been known since antiquity, but all attempts to find a method to solve third-degree equations (also called **cubics**) failed until the sixteenth century. In 1530, Niccolò FONTANA (1500 – 1557), nicknamed "Tartaglia" ("The stutterer") received two problems of cubic equations from Zuanne da Coi and announced that he had solved them, keeping the method used in secret.

Statue of Niccolò Fontana
Villa Borghese Gardens (Roma)
Photo: www.threesixty360.wordpress.com

However, Gerolamo CARDANO (1501 – 1576) convinced Tartaglia in 1539 to reveal him the method under the condition that he would never make it public or write a book about cubics without giving Tartaglia time to publish his own first. In 1545, breaking his promise, Cardano published the solution of the cubic in his book "Ars Magna" (at least, he mentioned that Tartaglia also knew it), although the method described was discovered, according to his words, by another mathematician called Ferro. Tartaglia, angry, challenged Cardano to a public competition to solve some equations (curious way to challenge someone, although perhaps better than the usual duel with guns of that time) that was not accepted. The challenge was taken up by a student of Cardano, Lodovico FERRARI (1522 – 1565), who won the competition and left Tartaglia without the money they had bet and, what is worse, without the prestige of being recognized as the inventor of the method that for centuries resisted the efforts of the mathematical community.

SOLUTION

We will deduce the complete method to solve any equation with complex coefficients of the type:

(15.1)
$$x^3 + px^2 + qx + r = 0$$

Note – Normally, the solution to the cubic is explained with a first step that consists of a change of variable $x = t - p/3$, which transforms the equation (15.1) into:

$$t^3 + \frac{3q - p^2}{3}t + \frac{2p^3 - 9pq + 27r}{27} = 0$$

As we can see, the quadratic term has disappeared. From there, we rewrite the equation as $t^3 + mt + n = 0$ and the Tartaglia method continues in a simpler way.

However, we are going to propose here a method that starts at the general equation, since this way we can observe the similarities with the solution to fourth-degree equations that we will see in another problem of the book, and we will also begin to understand what happens with fifth-degree equations that do not have a solution (also explained in detail in a problem of the second volume of this book).

Lagrange resolvent (for the solution of the cubic)

In his study of fifth-degree equations, the mathematician of French origin (although born in Turin) Joseph-Louis LAGRANGE found some functions, which he called resolvents, which helped him to find the solutions of the cubic and the quartic.

In the case of the cubic, let x_1, x_2, x_3 be the solutions of equation (15.1) and w_1, w_2 be the two cube roots of the unit different from 1 $[w_1 = (-1+i\sqrt{3})/2, w_2 = (-1-i\sqrt{3})/2]$. Since $w_1 = (w_2)^2$, we rename w_1 as w and w_2 as w^2.

The Lagrange resolvent is defined as the set of these two functions:

(15.2)
$$\begin{cases} A = x_1 + w \cdot x_2 + w^2 \cdot x_3 \\ B = x_1 + w^2 \cdot x_2 + w \cdot x_3 \end{cases}$$

Now we also define the functions that correspond to the previous ones cubed:

(15.3)
$$\begin{cases} a = A^3 = \left(x_1 + w \cdot x_2 + w^2 \cdot x_3\right)^3 \\ b = B^3 = \left(x_1 + w^2 \cdot x_2 + w \cdot x_3\right)^3 \end{cases}$$

With patience, let us check that both the values of $a + b$ and ab can be written as a function of the coefficients of equation (15.1). First, since x_1, x_2, x_3 are solutions of (15.1), we have the equation:

(15.4)
$$x^3 + px^2 + qx + r = (x - x_1) \cdot (x - x_2) \cdot (x - x_3)$$

from which we recover (equaling the coefficients that accompany each exponent) the famous Viete equations:

$$(15.5) \qquad \begin{cases} p = -(x_1 + x_2 + x_3) \\ q = x_1x_2 + x_1x_3 + x_2x_3 \\ r = -(x_1x_2x_3) \end{cases}$$

On the other hand, taking into account that $w + w^2 = -1$ and $w \cdot w^2 = 1$, we can expand the calculation of $a + b$ and ab:

$$a + b = \left(x_1 + w \cdot x_2 + w^2 \cdot x_3\right)^3 + \left(x_1 + w^2 \cdot x_2 + w \cdot x_3\right)^3 = \cdots = 2(x_1^3 + x_2^3 + x_3^3) + 12x_1x_2x_3 -$$
$$-3\left(x_1^2x_2 + x_1^2x_3 + x_2^2x_1 + x_2^2x_3 + x_3^2x_1 + x_3^2x_2\right)$$

$$ab = \left(x_1 + w \cdot x_2 + w^2 \cdot x_3\right)^3 \cdot \left(x_1 + w^2 \cdot x_2 + w \cdot x_3\right)^3 = \cdots = \left(x_1^2 + x_2^2 + x_3^2 - (x_1x_2 + x_1x_3 + x_2x_3)\right)^3$$

Substituting what we found in (15.5) in the previous equations:

$$a + b = 2(x_1^3 + x_2^3 + x_3^3) - 3\left(x_1^2x_2 + x_1^2x_3 + x_2^2x_1 + x_2^2x_3 + x_3^2x_1 + x_3^2x_2\right) - 12r$$
$$a \cdot b = (x_1^2 + x_2^2 + x_3^2 - q)^3$$

But, in addition, the reader can check the following equalities:

$$x_1^2 + x_2^2 + x_3^2 = (x_1 + x_2 + x_3)^2 - 2(x_1x_2 + x_1x_3 + x_2x_3) = p^2 - 2q$$

$$x_1^3 + x_2^3 + x_3^3 = (x_1 + x_2 + x_3)^3 - 3(x_1 + x_2 + x_3) \cdot (x_1x_2 + x_1x_3 + x_2x_3) + 3(x_1x_2x_3) = -p^3 + 3pq - 3r$$

$$\left(x_1^2x_2 + x_1^2x_3 + x_2^2x_1 + x_2^2x_3 + x_3^2x_1 + x_3^2x_2\right) = (x_1^2 + x_2^2 + x_3^2) \cdot (x_1 + x_2 + x_3) - (x_1^3 + x_2^3 + x_3^3) =$$
$$= (p^2 - 2q) \cdot (-p) - (-p^3 + 3pq - 3r) = -pq + 3r$$

so we can finally complete the computations of $a + b$ and ab:

$$a + b = 2 \cdot (-p^3 + 3pq - 3r) - 3 \cdot (-pq + 3r) - 12r$$
$$ab = \left((p^2 - 2q) - q\right)^3$$

or:

$$(15.6) \qquad \begin{cases} a + b & = -2p^3 + 9pq - 27r \\ ab & = \left(p^2 - 3q\right)^3 \end{cases}$$

Use of the resolvent to solve the cubic

The utility of the Lagrange resolvent is that we have managed to write two values ($a + b$ and ab) as a function of the coefficients of the original equation (15.1) and that these values are the coefficients of an equation of **second**-degree whose solutions are a and b. Similar to what we saw in (15.4), a second-degree equation with solutions a and b can be written as:

$$(15.7) \qquad x^2 - (a + b) \cdot x + ab = (x - a) \cdot (x - b)$$

Since we know how to solve second-degree equations, we can find the values of a and b as functions of the coefficients of the equation (15.7) ($a + b$ and ab). Since these values can be written, as we have seen in (15.6), as a function of the coefficients of equation (15.1), we deduce that we can find a and b as a function of p, q, r.

Once the values of a and b are found, we can infer the values of A and B (its cubic roots). And we ultimately have a linear system of three equations and three unknowns that allows us to finally find the solutions x_1, x_2, x_3:

$$(15.8) \qquad \begin{cases} A = x_1 + w \cdot x_2 + w^2 \cdot x_3 \\ B = x_1 + w^2 \cdot x_2 + w \cdot x_3 \\ -p = x_1 + x_2 + x_3 \end{cases}$$

The first two equations are those defined in (15.2), where now A and B are known values, while the third equation is the first Viete equation in (15.5), where p is also a known value.

Example: Resolution of a cubic equation

Let us take as an example the equation $x^3 - 107x^2 + 2131x - 2025 = 0$, which has integer solutions, so some readers will consider it to be simple (in fact, we could solve it by Ruffini's elementary method); the reason for using it is that this equation will appear when we solve an example of a quartic equation in a later chapter. However, the applied method will be perfectly understood and it is only necessary to bear in mind that the procedure would be identical for any other equation with complex numbers as coefficients.

First we calculate the values of $a + b$ and ab as found in (15.6), which gives values of:

$$a + b = -2 \cdot (-107)^3 + 9 \cdot (-107) \cdot 2131 - 27 \cdot (-2025) = 452608$$

$$ab = \left(107^2 - 3 \cdot 2131\right)^3 = 129247215616$$

Now we solve the equation $x^2 - (a + b) \cdot x + ab = 0$, which has a and b as solutions. Specifically, the equation $x^2 - 452608x + 129247215616$ has these solutions:

$$a = \frac{452608 + \sqrt{452608^2 - 4 \cdot 129247215616}}{2} \approx 226304 + 279345.15i$$

$$b = \frac{452608 - \sqrt{452608^2 - 4 \cdot 129247215616}}{2} \approx 226304 - 279345.15i$$

The next step is to calculate the cube roots of both values. For each value we have three possibilities:

$$A_1 = \sqrt[3]{a} \approx 68 + 20.78i \qquad A_2 = \sqrt[3]{a} \approx -52 + 48.50i \qquad A_3 = \sqrt[3]{a} \approx -16 - 69.28i$$
$$B_1 = \sqrt[3]{b} \approx 68 - 20.78i \qquad B_2 = \sqrt[3]{b} \approx -52 - 48.50i \qquad B_3 = \sqrt[3]{b} \approx -16 + 69.28i$$

By the definition of A and B in (15.2) and taking advantage of the calculations made in the previous section, it must be fulfilled that $AB = p^2 - 3q = 5056$. Therefore, not all pairs (A_i, B_j) lead us to a correct solution of (15.1). In this case, we only have solutions if we choose (A_1, B_1), (A_2, B_2) and (A_3, B_3).

If we choose (A_1, B_1) then we only need to solve the system of linear equations that we saw in (15.8):

$$68 + 20.78i - x_1 + (-1 + \sqrt{3})/2 \cdot x_2 + (-1 - \sqrt{3})/2 \cdot x_3$$

$$68 - 20.78i = x_1 + (-1 - \sqrt{3})/2 \cdot x_2 + (-1 + \sqrt{3})/2 \cdot x_3$$

$$107 = x_1 + x_2 + x_3$$

.. that gives us as solution the values $x_1 = 81$, $x_2 = 1$, $x_3 = 25$. In case that we had chosen (A_2, B_2), the values would have been the same, but in a different order ($x_1 = 1$, $x_2 = 25$, $x_3 = 81$), and (A_3, B_3) would have led us to $x_1 = 25$, $x_2 = 81$, $x_3 = 1$.

FINAL REMARKS

– When we try to solve the problem of the **quartic** (fourth-degree equation) in the future, we will also use a Lagrange resolvent. In that case, we will use three equations that are also a function of the solutions and that will later lead us to a lower grade equation (third-degree), similar to what we have seen in this problem.

As we now know how to solve third-degree equations, that will no longer be a problem and we will obtain a linear system of equations that will allow us to find the four solutions.

Having seen that the method worked for third and fourth-degree equations, Lagrange believed that it could be extended to fifth-degree equations and beyond. But despite his attempts, he could not find any resolvent functions that worked in this case. In the chapter dedicated to the **quintic** we will study in depth what happens so that the method does not work anymore.

- If we follow the method described for the coefficients of the equation (without particularizing them this time), we will find the generic "formula" to solve third-degree equations:

$$x = \frac{1}{3} \cdot \sqrt[3]{\frac{(2p^3 - 9pq + 27r) + \sqrt{(2p^3 - 9pq + 27r)^2 - 4(p^2 - 3q)^3}}{2}} +$$

$$+ \frac{1}{3} \cdot \sqrt[3]{\frac{(2p^3 - 9pq + 27r) - \sqrt{(2p^3 - 9pq + 27r)^2 - 4(p^2 - 3q)^3}}{2}} - \frac{p}{3}$$

Chapter 16

Sliding wheels

(Cardano – 1540)

PROBLEM

To describe the path (locus) followed by a marked point on a wheel that slides inside another wheel whose radius has double the length of the radius of the first wheel.

HISTORY

Gerolamo CARDANO was an Italian astrologer, mathematician and doctor (1501 – 1576) that passed to posterity mainly due to his formula for the resolution of third-degree equations that we have seen in the previous problem. As a doctor he came to treat the Pope, which did not prevent him from being accused of being a heretic and imprisoned for several months, until he publicly abjured the controversial texts he had published before (especially a horoscope of Jesus Christ).

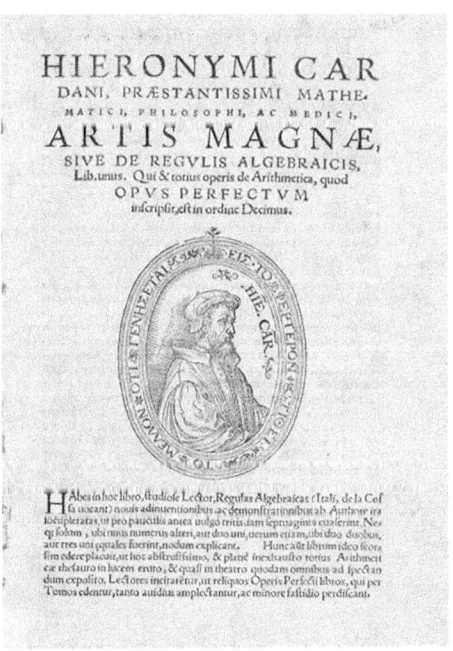

Old edition of "Ars Magna" (1545)
The work for which Cardano passed to posterity

Problems about geometric places (locus) were very common centuries ago, perhaps due to the need of creating new tools that, for example, transformed a circular movement into a linear one or vice versa. Throughout the book we will see some of them, all solved with elegant solutions that great mathematicians have left us as a legacy; for example, this problem of sliding wheels is solved by using only very simple geometric notions.

Let C_1 be the greater wheel (we can think of it as a circle) with radius r_1 and center in O_1, and let C_2 be the smaller wheel (another circle) with radius r_2 and center in O_2 which slides inside the first one. The conditions of the problem state that $r_1 = 2r_2$.

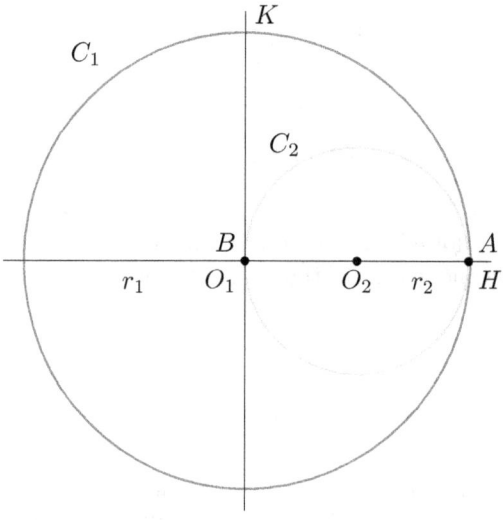

Figure 16.1

We will first see which path describes a diameter of C_2. Let A and H be the contact points of C_2 and C_1, respectively, at the beginning of the movement; let B be the opposite point of A in C_2 (so the segment $\overline{AB}$ is a diameter of C_2 and B is in O_1 at that moment); and let K be the point of C_1 that is at one end of the diameter that is perpendicular to $\overline{O_1H}$ (see figure 16.1).

Suppose now that after a certain amount of time, the wheel C_2 has moved (anticlockwise) an angle w (in radians), so that T is the new point of contact between both circles, X is the point of intersection between C_2 and $\overline{O_1H}$ and Y is the point of intersection between C_2 and $\overline{O_1K}$ (see figure 16.2).

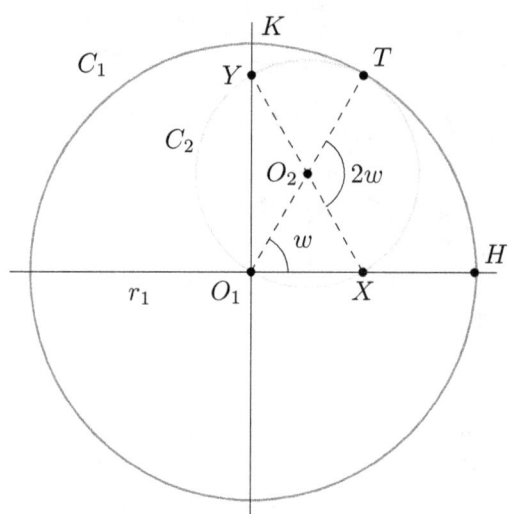

Figure 16.2

Using the property of inscribed angles (theorem 25.1), since $\widehat{XO_1Y}$ is a right angle it follows that $\overline{XY}$ is a diameter of C_2 and therefore the intersection between $\overline{XY}$ and O_1T is the center of C_2, i.e, O_2. Now, using the property of the central angle in a circle (theorem 25.1 again), it is inferred that the value of the angle $\widehat{XO_2T}$ is equal to $2w$.

The length of the arc TH can be calculated as follows: TH is an arc of angle w (in radians) and radius $r_1 = 2r_2$, so its length is equal to $2wr_2$. Besides, the arc TX has an angle of $2w$ and a radius whose value is equal to r_2, so its length is also $2wr_2$. What conclusion can we infer then? The point X is in fact our original point A of circle C_2 that originally occupied the same place as the fixed point H of circle C_2: as the circles move without sliding, that means that the lengths of the arcs XT and HT have the same value.

Now, if point X is the original point A and XY is a diameter of C_2, that means that point Y corresponds to the original point B, which has moved there.

Conclusion 1: "*The rotation of a circle inside another one with double radius causes the end points of any diameter of the inner circle to move along two fixed and orthogonal diameters of the greater circle*". In our example, point A moves along the diameter that passes through O_1 and H, while point B moves along the diameter that passes through O_1 and K (that is orthogonal to the first one).

It remains to determine what happens with any other point M of the smaller circle (not an end point of a diameter). In this case, the locus of point M is the same as that of a point in a segment whose end points A and B move along two sides of a right angle (sides $\overline{OK}$ and $\overline{OH}$). But, as we will see in problem 25 ("Sliding of a triangle"), this movement corresponds to a point of an ellipse.

Conclusion 2:" *The rotation of a circle inside another one with double radius causes any point of the inner circle other than the end point of a diameter to move along an ellipse*". In our example, point M moves along an ellipse (with center in O_1) with lengths of the semi-axes equal to the distances of M to both end points of the diameter that passes through it. Obviously, in case that M is the center O_2, the ellipse will become a circumference.

FINAL REMARKS

The study of the "sliding wheels" of Cardano and the locus that its points follow led to important advances in the construction of mechanical gears and high-speed printers, for example.

In the problem "The astroid" we will see again, in a rather surprising way, another example of sliding wheels, although with a different ratio between radii.

Chapter 17

Solution of the quartic

(Ferrari − 1545)

PROBLEM

To determine a general procedure to solve fourth-degree (quartic) equations with complex coefficients $x^4 + px^3 + qx^2 + rx + s = 0$.

HISTORY

Considering the great time lapse between the resolutions of second and third-degree equations, one might think that the next step from third to fourth-degree equations would take centuries. However, the difficulties found when solving the cubic made mathematicians of the time progress quickly and in fact Lodovico FERRARI, disciple of CARDANO and winner of the challenge against TARTAGLIA, published a correct method in the book "Ars Magna" (with Cardano as co-author). The procedure is based on the resolution of the cubic and it is believed that Ferrari discovered it even before knowing how to solve the cubic.

Statue of Lagrange in Turin
Photo: www.britannica.com

In this chapter we will not explain Ferrarri's method, but the one found by Joseph-Louis LA-GRANGE (1736 − 1813) making use of his famous resolvents, which we first saw in the solution of the cubic. This procedure, published in 1770, is significant because it brings together the solutions of the quadratic, the cubic and the quartic in a single mathematical principle. In spite of this, Lagrange did not succeed in extending it to the resolution of the quintic, but he laid the foundations for what is known as the GALOIS theory, a true revolution of algebra and usually the subject with which nowadays a mathematician ends his university studies.

SOLUTION

We are going to solve the fourth-degree equation (with complex coefficients):

$$(17.1) \qquad x^4 + px^3 + qx^2 + rx + s = 0$$

that has values x_1, x_2, x_3 and x_4 as solutions. The resolution method will be very similar to the one we apply for third-degree equations, starting by defining the Lagrange resolvent.

Lagrange resolvent (for the resolution of the quartic)

We define the Lagrange resolvent as the following three functions:

$$(17.2) \qquad \begin{cases} A = x_1 + x_2 - x_3 - x_4 \\ B = x_1 - x_2 + x_3 - x_4 \\ C = x_1 - x_2 - x_3 + x_4 \end{cases}$$

We now define the functions that consists of the previous ones squared:

$$(17.3) \qquad \begin{cases} a = A^2 = (x_1 + x_2 - x_3 - x_4)^2 \\ b = B^2 = (x_1 - x_2 + x_3 - x_4)^2 \\ c = C^2 = (x_1 - x_2 - x_3 + x_4)^2 \end{cases}$$

Each value $a + b + c$, $ab + ac + bc$ and abc can be written as a function of the coefficients p, q, r, s of equation (17.1). First, since x_1, x_2, x_3, x_4 are solutions of (17.1), we have the equality:

$$(17.4) \qquad x^4 + px^3 + qx^2 + rx + s = (x - x_1) \cdot (x - x_2) \cdot (x - x_3) \cdot (x - x_4)$$

from which we infer (by equaling the coefficients of the appropriate power of x of each side) the Viete equations:

$$(17.5) \qquad \begin{cases} p = -(x_1 + x_2 + x_3 + x_4) \\ q = (x_1x_2 + x_1x_3 + x_1x_4 + x_2x_3 + x_2x_4 + x_3x_4) \\ r = -(x_1x_2x_3 + x_1x_2x_4 + x_1x_3x_4 + x_2x_3x_4) \\ s = (x_1x_2x_3x_4) \end{cases}$$

Similarly to what happened in the case of the cubic, we can now calculate $a + b + c$, $ab + ac + bc$ and abc as a function of the coefficients p, q, r, s. The calculations are laborious, but the interested reader will have no problems checking the following equalities:

$$(17.6) \qquad \begin{aligned} a + b + c &= 3p^2 - 8q \\ ab + ac + bc &= 3p^4 - 16p^2q + 16pr + 16q^2 - 64s \\ abc &= (p^3 - 4pq + 8r)^2 \end{aligned}$$

Use of the resolvent to solve the quartic

With the Lagrange resolvent we have managed to write three values $(a + b + c$, $ab + ac + bc$ and $abc)$ as functions of the coefficients of the original equation (17.1) and these values are the coefficients of a third-degree equation whose solutions are a, b and c. In fact, similarly to those seen in (17.4), a third-degree equation with solutions a, b and c can be written as:

$$(17.7) \qquad x^3 - (a + b + c)x^2 + (ab + ac + bc)x + abc = (x - a) \cdot (x - b) \cdot (x - c)$$

Since we know how to solve third-degree equations, we can find values a, b and c as functions of the coefficients of equation (17.7) (known values $a + b + c$, $ab + ac + bc$ and abc). Since these values can be written, as we have seen in (17.6), as a function of the coefficients of equation (17.1), we deduce that we can find a, b and c as functions of p, q, r, s.

Once the values of a, b and c have been found, we infer the values of A, B and C (its square roots). And we finally have a linear system of four equations and four unknowns that allows us to find the solutions x_1, x_2, x_3, x_4:

$$(17.8) \qquad \begin{cases} A = x_1 + x_2 - x_3 - x_4 \\ B = x_1 - x_2 + x_3 - x_4 \\ C = x_1 - x_2 - x_3 + x_4 \\ -p = x_1 + x_2 + x_3 + x_4 \end{cases}$$

The first three equations are those defined in (17.2), where A, B and C are now known numbers, while the fourth one is the first Viete equation in (17.5), where p is also a known value.

Example: Resolution of a quartic equation

Let us take as an example the equation $x^4 - 11x^3 + 32x^2 - 4x - 48 = 0$, which has very easy integer solutions but allows us to understand the method (identical for any other fourth-degree equation with complex numbers as coefficients). First we calculate the values of $a + b + c$, $ab + ac + bc$ and abc as found in (17.6), which lead us to the following values:

$$a + b + c = 3 \cdot (11)^2 - 8 \cdot (32) = 107$$
$$ab + ac + bc = 3 \cdot (11)^4 - 16 \cdot (-11^2) \cdot 32 + 16 \cdot (-11) \cdot (-4) + 16 \cdot (32)^2 - 64 \cdot (-48) = 2131$$
$$abc = (11^3 - 4 \cdot 11 \cdot 32 + 8 \cdot (-4))^2 = 2025$$

Now we solve the equation $x^9 - (a + b + c)x^2 + (ab + ac + bc)x + abc = 0$, which has a, b and c as solutions. The equation that we get is $x^3 - 107x^2 + 2131x - 2025 = 0$, precisely the example that we discussed in the chapter of third-degree equations, whose solutions were:

$$a = 81 \quad ; \quad b = 1 \quad ; \quad c = 25$$

The next step is to calculate the square roots of the previous values. For each value we have two possibilities:

$$A_1 = \sqrt{a} = 9 \quad ; \quad B_1 = \sqrt{b} = 1 \quad ; \quad C_1 = \sqrt{c} = 5$$
$$A_2 = \sqrt{a} = -9 \quad ; \quad B_2 = \sqrt{b} = -1 \quad ; \quad C_2 = \sqrt{c} = -5$$

Bearing in mind the definition of A, B and C in (17.2) and taking advantage of the calculations made in the previous section, it must be fulfilled that $ABC = -p^3 + 4pq - 8r = -45$, so not all the possibilities of (A_i, B_j, C_k) lead us to a correct solution of (17.1). In this case, we only have solutions if we choose (A_1, B_1, C_2), (A_1, B_2, C_1), (A_2, B_1, C_1) and (A_2, B_2, C_2). If we choose (A_1, B_1, C_2) then we only need to solve the system of linear equations (17.8):

$$9 = x_1 + x_2 - x_3 - x_4$$
$$1 = x_1 - x_2 + x_3 - x_4$$
$$-5 = x_1 - x_2 - x_3 + x_4$$
$$11 = x_1 + x_2 + x_3 + x_4$$

.. which yields the solution of values $x_1 = 4$, $x_2 = 6$, $x_3 = 2$, $x_4 = -1$. If we choose (A_1, B_2, C_1), the values found are the same but in a different order ($x_1 = 6$, $x_2 = 4$, $x_3 = -1$, $x_4 = 2$); if we choose (A_2, B_1, C_1) we obtain ($x_1 = 2$, $x_2 = -1$, $x_3 = 4$, $x_4 = 6$); and if we choose (A_2, B_2, C_2) the solution is ($x_1 = -1$, $x_2 = 2$, $x_3 = 6$, $x_4 = 4$).

FINAL REMARKS

The fact that $a + b + c$ (and $ab + ac + bc$ or abc) can be written as a polynomial of the coefficients of the original equation (equation (17.6)) is **NOT** a curious coincidence. $a + b + c$ is an example of what we call a **symmetric polynomial** in four variables. A symmetric polynomial is a polynomial that does not change if we vary the order of its variables in the definition.

$$a + b + c = (x_1 + x_2 - x_3 - x_4)^2 + (x_1 - x_2 + x_3 - x_4)^2 + (x_1 - x_2 - x_3 + x_4)^2$$

That is, the role of the variables x_1, x_2, x_3, x_4 is symmetric in the definition of $a + b + c$.

There is a beautiful theorem that shows that every symmetric polynomial of n variables can be written as a polynomial of the so-called **elementary symmetric polynomials** of n variables. But which are the elementary symmetric polynomials by definition? For $n = 4$ variables, the elementary symmetric polynomials are:

$$\begin{cases} x_1 + x_2 + x_3 + x_4 \\ x_1 x_2 + x_1 x_3 + x_1 x_4 + x_2 x_3 + x_2 x_4 + x_3 x_4 \\ x_1 x_2 x_3 + x_1 x_2 x_4 + x_1 x_3 x_4 + x_2 x_3 x_4 \\ x_1 x_2 x_3 x_4 \end{cases}$$

which are the same (maybe with a different sign) as the coefficients of equation (17.1). In general, the elementary symmetric polynomials of n variables also coincide with Viete's formulas for n-degree equations.

Chapter 18

A terrestrial conformal map

(Mercator – 1569)

PROBLEM

To draw a terrestrial map in which every angle is preserved (i.e., a conformal map) and the equator and meridians are straight lines.

HISTORY

When we want to draw a map of the surface of Earth, there is always a problem: there is not a "perfect" way to convey the original information (on a sphere) into a map (on a plane). There are maps that preserve areas (regions of Earth that have the same area also have equal areas on the map), others that preserve distances, others that preserve directions, ... but none of them can keep everything preserved.

Each way of representing Earth onto a map is known as a cartographic projection and the most common ones are usually named after the corresponding designer. The Mercator's projection, which is equally important for the geographical and nautical sciences, was developed by the German (although born in Flanders) cartographer Gerhard KREMER (1512 – 1594), also known as Mercator.

Picture of Mercator
Frans Hogenberg (1574)

We will define the Mercator's projection as he devised it and we will see that it is a solution to our problem, taking into account that a map is conformal by definition when it preserves angles, i.e., every angle on the map is equal to its original angle on the sphere. The price to pay in order to preserve angles will be not being able to preserve areas, so that means that countries that are close to the poles are bigger on the map than those countries that are near the equator.

SOLUTION

Parametrization of the sphere

In Mathematics, one way to define a surface is via a parameterization using two variables. That is, we take two variables (for example, u and v) that can vary within a range of known values and for each pair of them we define a point in R^3 by means of a formula where each coordinate depends on u and v.

The first example of a parameterization of a surface that is often used is the sphere of radius equal to 1 with center in O. We choose a random circle of the sphere called **equator** and a circle that is perpendicular to it (that is, the plane that contains that circle and the plane that contains the equator are perpendicular) called **main meridian**; the rest of the circles which are perpendicular to the equator will also be called meridians.

Let A be a point of the sphere. We define u as the angle (in radians) between the plane which contains the main meridian and the plane which contains the meridian where A is (this angle is called **longitude**), and we define v as the angle (in radians) between the vector $\overrightarrow{OA}$ and the plane that contains the equator (this angle is called **latitude**). Then, any point of the sphere A can be defined with the appropriate values u and v in the following way:

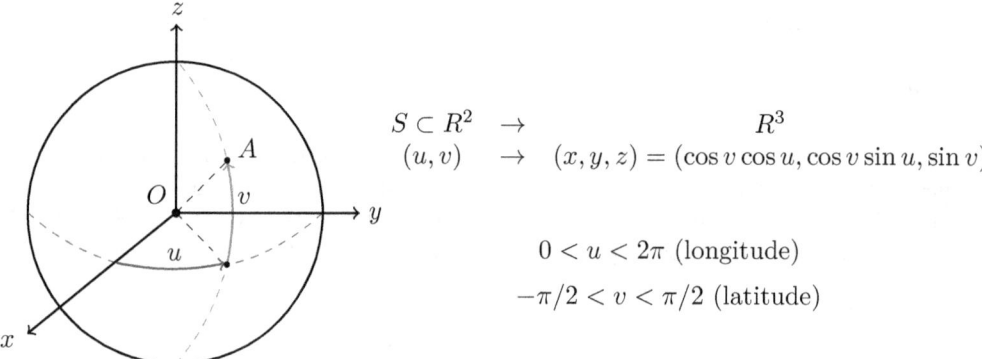

$$
\begin{array}{ccc}
S \subset R^2 & \to & R^3 \\
(u, v) & \to & (x, y, z) = (\cos v \cos u, \cos v \sin u, \sin v)
\end{array}
$$

$$0 < u < 2\pi \ \text{(longitude)}$$

$$-\pi/2 < v < \pi/2 \ \text{(latitude)}$$

Figure 18.1

Obviously, the reader can check that the equation $x^2 + y^2 + z^2 = 1$ is satisfied for each point of the sphere, regardless of the values of u and v.

Although this is the most well-known parameterization of the sphere, it is not the only one; in fact, we can think of many possible parameterizations. Due to reasons that we will be later explained, let us now study one of them, which we call parameterization T and is shown in figure 18.2, that is slightly different from the one we have seen before.

Notice that we have changed the names of the variables (to denote them by Greek letters that are often used in Mathematics to define angles) and that we have used the variable **colatitude** (with the North Pole as the origin and $[0, \pi]$ as range) instead of using the well-known latitude (with the equator as the origin and $[-\pi/2, \pi/2]$ as range).

The parameterization T transforms a point on a plane (two variables) into a point on the sphere (three variables), while the inverse parameterization T^{-1} transforms a point of the sphere into a point on the plane (it is therefore a cartographic projection - that is, a **map** -, as we defined in the introduction). The map projection T^{-1} represents the equator as a line ($\theta = \pi/2$) and the meridians as lines that are perpendicular to it ($\phi = $ constant), but the reader can check that this map is **not** a conformal application (it does not preserve angles).

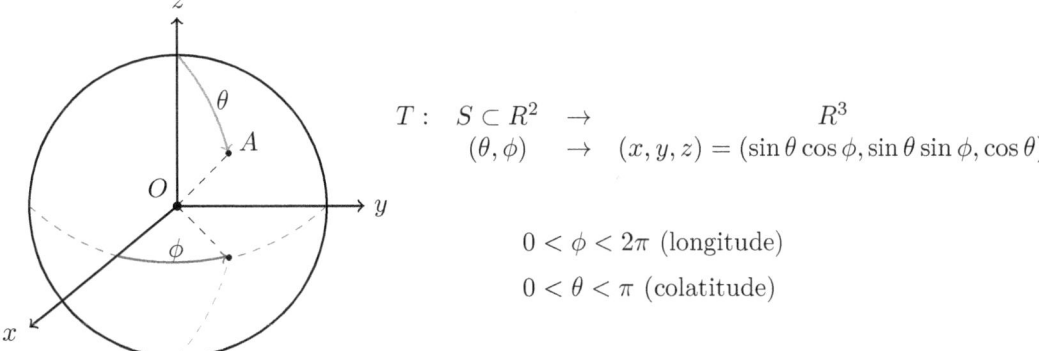

$$T: \quad S \subset R^2 \quad \to \qquad\qquad\qquad R^3$$
$$(\theta, \phi) \quad \to \quad (x, y, z) = (\sin\theta\cos\phi, \sin\theta\sin\phi, \cos\theta)$$

$$0 < \phi < 2\pi \ (\text{longitude})$$
$$0 < \theta < \pi \ (\text{colatitude})$$

Figure 18.2

This first attempt to find a conformal map has failed. To obtain the solution we will think about the problem from another point of view.

Loxodromes

A **loxodrome** is a curve on the terrestrial surface that cuts all the meridians with the same angle. Whenever a ship does not change its course, it is sailing on a loxodrome. The angle k formed by the loxodrome with the meridians is called the **azimuth** of the route.

Similarly to what we have seen with surfaces, one way to define a trajectory is through a parameterization; in this case, only one variable is necessary (since we are defining a one-dimensional entity), unlike the two variables we need for a surface (a two-dimensional entity). Normally, for the sake of convenience and a better understanding, time t is used as the parameter so that for each value of t the parameterization defines a point (x, y, z).

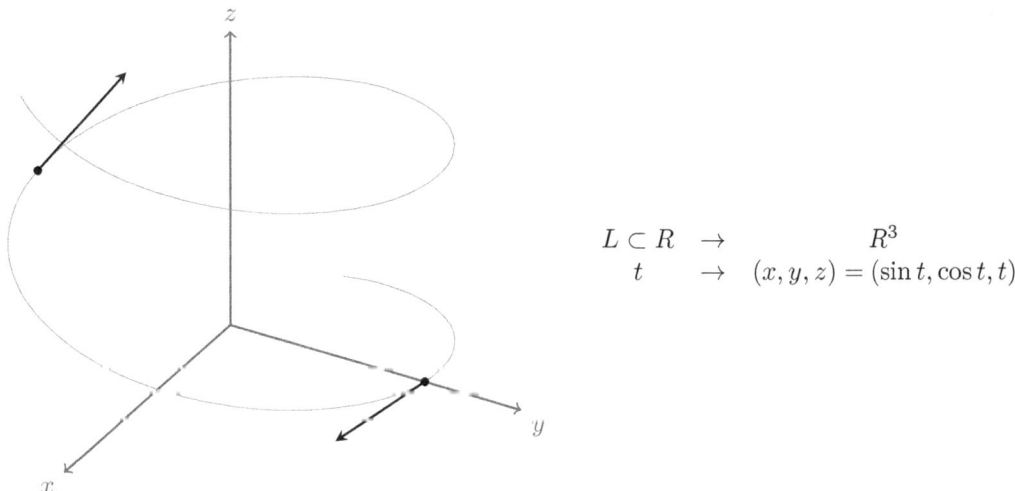

$$L \subset R \quad \to \qquad\qquad R^3$$
$$t \quad \to \quad (x, y, z) = (\sin t, \cos t, t)$$

Figure 18.3

For example, the parameterization in figure 18.3 defines the trajectory of an helix: for $t = 0$ we have that the imaginary point that follows the trajectory is at $(0, 1, 0)$, and for $t = \pi$ is at $(0, -1, \pi)$; that is, for each value of time (in the defined range) we know where the point is. The path followed by the point is defined thanks to the parameter t (hence the concept of parameterization).

Although we will not do it here, it can be shown that the derivative of the parameterization (component to component) at a point indicates the direction of the tangent to the trajectory at that point. In the previous example, the derivative of the parameterization of the helix is

$(\cos t, -\sin t, 1)$ which means that, for example, the tangent to the trajectory at $t = 0$ has direction $(1, 0, 1)$ and the tangent to the trajectory at $t = \pi$ has direction $(-1, 0, 1)$, as we can see in figure 18.3.

Now that this preamble was explained we will try to find a parameterization of a loxodrome. Let us start with a parameterization of any trajectory that runs along the surface of the sphere:

$$(18.1) \qquad \alpha(t) = (\sin \theta(t) \cos \phi(t), \sin \theta(t) \sin \phi(t), \cos \theta(t))$$

Notice that we have used the parameterization of the sphere of figure 18.2 and we have just written variables θ and ϕ as functions of the (same) parameter t. That is, we force the two variables θ and ϕ to stop being independent and to start being defined by a single parameter t. This implies that we have a single-dimensional entity (a trajectory), but we can be sure that this trajectory always occurs on the surface of the sphere of radius equal to 1.

In the generic equation (18.1), in order to find a a parametrization of the loxodrome we have to impose now the condition that the trajectory always forms the same angle with the meridians; this is not easy, but we will do it right away.

The parameterization of a meridian must satisfy that its longitude is constant, that is, $\theta(t) = C$, which implies that both $\cos \theta(t)$ and $\sin \theta(t)$ are constant values (let A and B be these two values, respectively; it is satisfied that $A^2 + B^2 = 1$). Therefore, a meridian has the following parametrization (the values of A and B vary according to the chosen meridian):

$$(18.2) \qquad \gamma(t) = (A \sin \theta(t), B \sin \theta(t), \cos \theta(t))$$

Now we have to differentiate both parameterizations in order to - as we have explained before - calculate the tangent vectors to the trajectories and then calculate the angle formed by both of them; that angle must always be the same by definition of loxodrome.

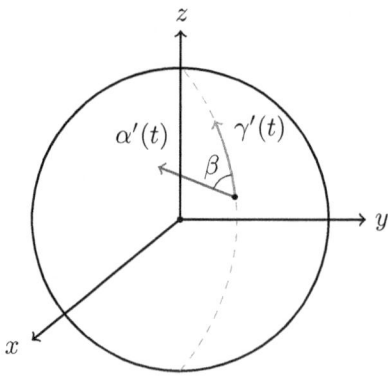

Figure 18.4

Let us first differentiate parameterizations (18.1) and (18.2):

$$\alpha'(t) = (\cos \theta(t) \cos \phi(t) \cdot \theta'(t) - \sin \theta(t) \sin \phi(t) \cdot \phi'(t),$$
$$\cos \theta(t) \sin \phi(t) \cdot \theta'(t) + \sin \theta(t) \cos \phi(t) \cdot \phi'(t), -\sin \theta(t) \cdot \theta'(t))$$
$$\gamma'(t) = (A \cos \theta(t) \cdot \theta'(t), -B \sin \theta(t) \cdot \theta'(t), -\sin \theta(t) \cdot \theta'(t))$$

106

and now calculate its values at a certain time t_1 (we assume that $\cos \theta(t_1) = A$ and $\sin \theta(t_1) = B$):

$$\alpha'(t_1) = (A \cos \phi(t_1) \cdot \theta'(t_1) - B \sin \phi(t_1) \cdot \phi'(t_1),$$
$$A \sin \phi(t_1) \cdot \theta'(t_1) + B \cos \phi(t_1) \cdot \phi'(t_1), -B \cdot \theta'(t_1))$$
$$\gamma'(t_1) = (A^2 \cdot \theta'(t_1), -B^2 \cdot \theta'(t_1), -B \cdot \theta'(t_1))$$

Finally, the angle β between both vectors at time t_1 can be calculated with the following formula, as we will see in proposition 46.1 (in the second part of this book):

$$\cos \beta = \frac{\langle \alpha'(t_1), \gamma'(t_1) \rangle}{|\alpha'(t_1)| \cdot |\gamma'(t_1)|}$$

where $\langle *, * \rangle$ is the scalar product of two vectors and $|*|$ is the norm of the vector. The calculations are laborious, but the reader can verify that the final result is:

$$\cos \beta = \frac{\theta'(t_1)}{\sqrt{(\theta'(t_1))^2 + \sin^2 \theta(t_1) \cdot (\phi'(t_1))^2}}$$

Now, by definition of a loxodrome, we have to impose the condition that the angle β has a constant value for any time t. So the equation of the loxodrome fulfills the following differential equation in which, for convenience, we write θ' and ϕ' instead of $\theta'(t)$ and $\phi'(t)$:

$$(18.3) \qquad \cos \beta = \frac{\theta'}{\sqrt{(\theta')^2 + \sin^2 \theta \cdot (\phi')^2}}$$

To solve (18.3) we must first notice that it is equivalent to write (applying that $\sin^2 \beta + \cos^2 \beta = 1$):

$$(18.4) \qquad \sin \beta = \frac{\sin \theta \cdot \phi'}{\sqrt{(\theta')^2 + \sin^2 \theta \cdot (\phi')^2}}$$

Now, dividing (18.4) by (18.3):

$$\tan \beta = \frac{\pm \sin \theta \cdot \phi'}{\theta'} \qquad \Rightarrow \qquad \frac{\theta'}{\sin \theta} = \pm \frac{\phi'}{\tan \beta}$$

The last equation is an ordinary differential equation with separated variables that can be solved by integration of both sides:

$$(18.5) \qquad \ln\left(\tan\left(\frac{\theta}{2}\right)\right) = \pm \frac{\phi + c}{\tan \beta}$$

This is the equation that defines a loxodrome, which includes two variables of the parameterization (longitude and colatitude), constant azimuth β and a constant value c that can be calculated if we know any point of the trajectory. The sign $\pm$ depends on whether we are heading North or South (in case of going with a constant latitude, the equation is trivial).

Note: Equation (18.5) explains the need to choose the colatitude variable (figure 18.2, where $0 < \phi < \pi$) instead of the latitude variable (figure 18.1, where $-\pi/2 < \phi < \pi/2$), since this way we make sure that $\tan(\phi/2)$ is always a positive value and, therefore, we can calculate $\ln(\tan(\phi/2))$.

Mercator's projection

Once the equation of a loxodrome (18.5) is established, we can proceed as follows:

We are looking for a projection in which the angles are preserved. We can look for one where meridians and loxodromes are both transformed into straight lines on the map to achieve that a loxodrome cuts all meridians with the same angle.

In order to get loxodromes are transformed into straight lines in the projection, one way to do it will be with the following change of variable:

$$(18.6) \qquad \begin{cases} u = -\ln\left(\tan\left(\frac{\theta}{2}\right)\right) \\ v = \phi \end{cases}$$

With this change of variable, meridians (ϕ constant) become straight lines on the map (v constant), but also loxodromes, since the equation (18.5) found in the previous section remains as:

$$u = \mp\frac{v + c}{\tan\beta}$$

and that defines a line in the variables (u, v). The negative sign in the definition of u is just because we want that points of the Northern Hemisphere ($0 < \theta < \pi/2$) meet the condition $u > 0$, while those of the Southern Hemisphere ($\pi/2 < \theta < \pi$) meet the condition $u < 0$.

Therefore, the function (18.6) is the projection we are looking for because it transforms a point on the sphere (expressed as a function of the longitude and colatitude) into a point on the plane (i.e., Mercator's map) in a way that the values of angles are preserved (an angle on the sphere is always equal to the angle between two well-chosen loxodromes; when loxodromes are transformed into lines on the map their angles with the meridians are preserved and, as a consequence, the angle between them).

FINAL REMARKS

- Although it is not intuitive at all, we must note the difference between following a fixed course (loxodrome) and not varying the rudder of the ship (in this case the trajectory is a maximum circle of the sphere). The first man to realize this was the Portuguese scientist Pedro Nunes (1492 – 1577): any sailor that lived before him was convinced that if a ship sails on a fixed course then its trajectory would be in a maximum circle that would take it to the same point of departure after a round to the world (if there is no land in between). Nunes was the first man to warn about this error, ensuring that a ship with a fixed course (loxodrome) would follow an spiral trajectory that would spiral closer and closer to one of the poles of Earth without ever reaching it.

- Mercator's projection was an attempt to help ships in their navigation, allowing to calculate fixed bearings on the map with accuracy. Although it is very useful for navigation, other projections are used in other fields, such as the Lambert conic projection (for air navigation) or the Winkel-Tripel projection (used by the National Geographic Society).

- For interested readers, there is a simple way to find the inverse function of Mercator's projection, i.e., the function that transforms a point of the Mercator map into its corresponding one on the terrestrial sphere. In order to find it, we apply the trigonometric formula of the half angle in equations (18.6):

$$e^{-u} = \tan\left(\frac{\theta}{2}\right) = \frac{1 + \sec\theta}{\tan\theta} = \frac{1 + cos\theta}{\sin\theta} \quad \text{and, then,} \quad e^{u} = \frac{\sin\theta}{1 + \cos\theta}$$

Adding and subtracting both equations:

$$e^{u} + e^{-u} = \frac{\sin\theta}{1 + \cos\theta} + \frac{1 + cos\theta}{\sin\theta} = \frac{1 + 2\cos\theta + \cos^2\theta + \sin^2\theta}{\sin\theta \cdot (1 + \cos\theta)} =$$
$$= \frac{2 + 2\cos\theta}{\sin\theta \cdot (1 + \cos\theta)} = \frac{2}{\sin\theta}$$

$$e^{u} - e^{-u} = \frac{\sin\theta}{1 + \cos\theta} - \frac{1 + cos\theta}{\sin\theta} = \frac{\sin^2\theta - 1 - 2\cos\theta - \cos^2\theta}{\sin\theta \cdot (1 + \cos\theta)} =$$
$$= \frac{-2\cos\theta - 2\cos^2\theta}{\sin\theta \cdot (1 + \cos\theta)} = \frac{-2}{\tan\theta}$$

That is:

$$\sin\theta = \frac{2}{e^{u} + e^{-u}} \quad \text{and} \quad \cos\theta = \frac{-e^{u} + e^{-u}}{e^{u} + e^{-u}}$$

Therefore, replacing the previous formulas in the parameterization x of the sphere proposed in figure 18.1, the inverse function of Mercator's projection is:

$$y: \quad V \subset R^2 \quad \to \qquad\qquad\qquad R^3$$
$$(u, v) \quad \to \quad \left(\frac{2}{e^{u}+e^{-u}} \cdot \cos v, \frac{2}{e^{u}+e^{-u}} \cdot \sin v, \frac{-e^{u}+e^{-u}}{e^{u}+e^{-u}}\right)$$

$$-\infty < u < \infty \text{ ("exaggerated" latitude)}$$
$$0 < v < 2\pi \text{ (longitude)}$$

Chapter 19

Length of a loxodrome

(Harriot − 1590)

PROBLEM

To determine the length of the loxodrome that joins two points on the surface of Earth and to check that it is not the minimum length between these two points.

HISTORY

As we saw in the previous problem, Mercator's projection transforms a loxodrome (whose name that comes from the dutch Willebrord Snell) into a line on the map, which makes it very simple to calculate distances between two points joined by it. The first man who made the calculations (it is believed that he did so by using something similar to the integral calculus discovered centuries later by Newton and Leibniz) was the English mathematician Thomas HARRIOT (1560 − 1621), although we will apply more advanced techniques here.

Portrait of Thomas Harriot
Unknown author (1602)

As Nunes had already observed, Harriot proved that the distance following a loxodrome was not the minimum between two points, which was a very important information to take into account by the sailors of that time: in long trips the best option is not to take a fixed course (loxodrome), but to follow the so-called maximum circle. Today this advice is well known, and aircrafts use it as we can check on the screens of our seats on intercontinental flights (the course followed by the plane appears as a curved line on the screen).

We will explain how to perform both calculations (distances in a loxodrome and in a maximum circle) and we will provide a clarifying example.

SOLUTION

Length of a parameterized regular curve

First, let us see how the length of a curve on a surface is calculated, in this case on a sphere. According to the theory of curves and surfaces that interested readers can check, the length between points t_0 and t of a regular curve parameterized by $\alpha(t)$ can be calculated as:

$$(19.1) \qquad L = \int_{t_0}^{t} |\alpha'(t)| \cdot dt$$

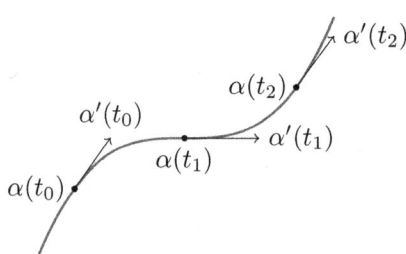

Figure 19.1

The proof of (19.1) is not simple, but we can understand the idea behind it: we approximate "parts" of the curve ($\alpha(t)$) starting each one at a point by its tangent vectors ($\alpha'(t)$) on those points. We then calculate the length of those vectors ($|\alpha'(t)|$) and add them all; the closer the points considered are, the better our approximation is and in the limit the sum is calculated with an integral.

Consider that Earth is a sphere of radius equal to 1 and that $\alpha(t)$ is a curve that is on its surface. Then, considering the parameterization that we saw in formula 18.1:

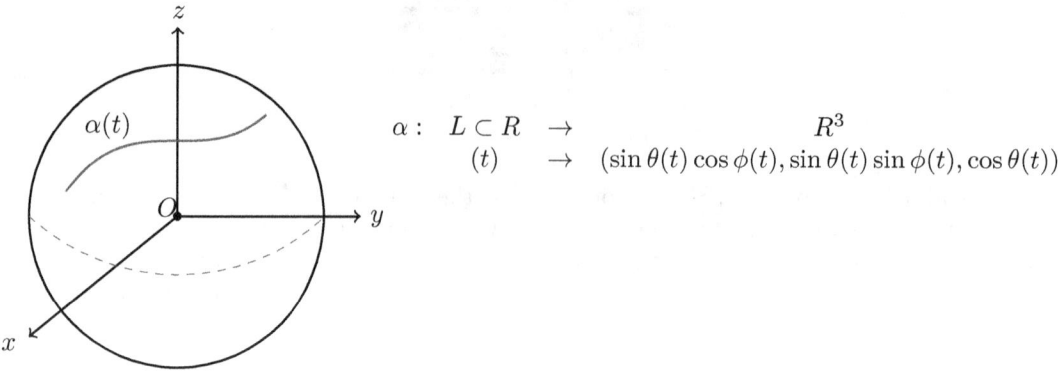

Figure 19.2

we have that a tangent vector is:

$$\alpha'(t) = (\,\cos\theta(t)\cos\phi(t)\theta'(t) - \sin\theta(t)\sin\phi(t)\phi'(t),$$
$$\cos\theta(t)\sin\phi(t)\theta'(t) + \sin\theta(t)\cos\phi(t)\phi'(t), -sin\theta(t)\theta'(t))$$

112

Let us now calculate its length, with the help of the scalar product:

$$\begin{aligned}
|\alpha'(t)|^2 = <\alpha'(t), \alpha'(t)> &= \cos^2\theta(t)\cos^2\phi(t)(\theta'(t))^2 + \sin^2\theta(t)\sin^2\phi(t)(\phi'(t))^2 - \\
&- 2\cos\theta(t)\cos\phi(t)\theta'(t)\sin\theta(t)\sin\phi(t)\phi'(t) + \\
&+ \cos^2\theta(t)\sin^2\phi(t)(\phi'(t))^2 + \sin^2\theta(t)\cos^2\phi(t)(\phi'(t))^2 - \\
&+ 2\cos\theta(t)\cos\phi(t)\theta'(t)\sin\theta(t)\sin\phi(t)\phi'(t) + \\
&+ \sin^2\theta(t)\cdot(\theta'(t))^2 = (\theta'(t))^2 + \sin^2\theta(t)\cdot(\phi'(t))^2
\end{aligned}$$

$$(19.2) \qquad |\alpha'(t)| = \sqrt{(\theta'(t))^2 + \sin^2\theta(t)\cdot(\phi'(t))^2}$$

Equation (19.2) is true for any curve contained on Earth's surface but for a **loxodrome** we saw in the previous problem that the following condition it is also fulfilled:

$$\cos\beta = \frac{\theta'(t)}{\sqrt{(\theta'(t))^2 + \sin^2\theta(t)\cdot(\phi'(t))^2}}$$

so, replacing what we found in (19.2):

$$(19.3) \qquad |\alpha'(t)| = \frac{\theta'(t)}{\cos\beta}$$

Therefore, the equation (19.3) is true only for a loxodrome, which we can use in the formula (19.1) for the calculation of the length of the curve:

$$(19.4) \qquad L = \int_{t_0}^{t} |\alpha'(t)|\cdot dt = \int_{t_0}^{t} \frac{\theta'(t)}{\cos\beta}\cdot dt = \int_{\theta_0}^{\theta} \frac{d\theta}{\cos\beta} = \frac{\theta - \theta_0}{\cos\beta}$$

where θ_0 and θ are, respectively, the colatitudes of the initial and final point, and β is the angle between the loxodrome and the meridians.

That is, the length of a curve on Earth's surface is very easy to calculate if the curve turns out to be a loxodrome. It is only necessary to calculate the azimuth (value of β) and, in order to achieve this, it is better to take advantage of the fact that loxodromes are straight lines on Mercator's map (as we saw in the previous problem) to calculate it there.

Example

Calculate the length of the loxodrome between Valdivia (Chile) and Yokohama (Japan).

Values for Valdivia: $\phi_1 = 286.582° = 5.002$ radians (longitude)
$\theta_1 = 129.885° = 2.267$ radians (colatitude)

Values for Yokohama: $\phi_2 = 139.653° = 2.437$ radians (longitude)
$\theta_2 = 054.557° = 0.952$ radians (colatitude)

113

First, we look for these two points in the map, by means of Mercator's projection (again, using the results of the previous problem):

$$u_1 = -\ln\left(\tan\left(\frac{\theta_1}{2}\right)\right) = -0.7604 \qquad v_1 = \phi_1 = 5.002$$

$$u_2 = -\ln\left(\tan\left(\frac{\theta_2}{2}\right)\right) = -0.6623 \qquad v_2 = \phi_2 = 2.437$$

On the map, we easily calculate the angle between the line joining both points and a meridian:

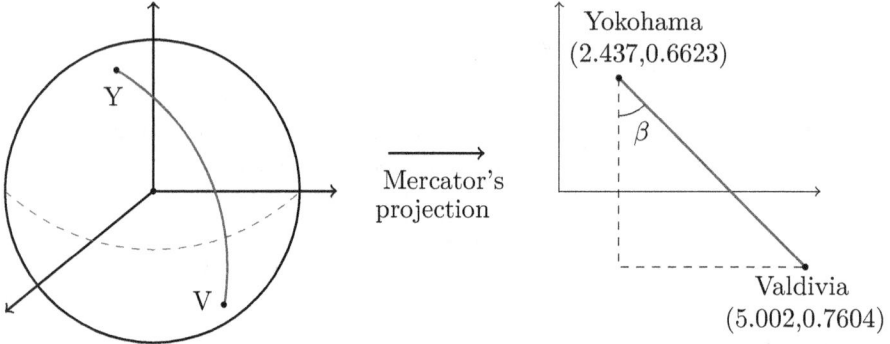

Figure 19.3

Therefore,

$$\tan\beta = \frac{5.002 - 2.437}{0.7604 + 0.6623} = 1.803 \qquad \Rightarrow \qquad \beta = 1.064 \text{ radians } (60.98°)$$

Since the map is conformal, the calculated angle is equal to the real one on Earth's surface, so we can now calculate the desired length using formula (19.4), which is valid only for loxodromes:

$$L = \frac{\theta - \theta_0}{\cos\beta} = \frac{2.267 - 0.952}{\cos 1.064} = 2.709$$

All these calculations have been made taking the radius of Earth to be equal to 1. But the radius of Earth is approximately 6367 km, so all distances must now be multiplied by it:

Length of the loxodrome Valdivia-Yokohama $= 2.709 \cdot 6367$ Km $= 17247$ Km

Calculate the minimum surface length between Valdivia and Yokohama.

The shortest path between two points of a sphere is known as the **geodesic** and it can be proved, using theory of curves and surfaces, that it is the line that results from the intersection between the sphere and the plane that contains both points and the center of Earth (this plane always exists and it is unique, except in the case that both points are opposite in the globe: in such case there are infinite planes and for each of them there is a geodesic).

To calculate the distance it is enough to calculate the angle between the two vectors that join the center with our two points and multiply this value (in radians) by the terrestrial radius:

$$a_1 = (\sin 2.267 \cdot \cos 5.002, \sin 2.267 \cdot \sin 5.002, \cos 2.267) = (0.2191, -0.7353, -0.6413)$$

$$a_2 = (\sin 0.952 \cdot \cos 2.437, \sin 0.952 \cdot \sin 2.437, \cos 0.952) = (-0.6206, 0.5276, 0.5800)$$

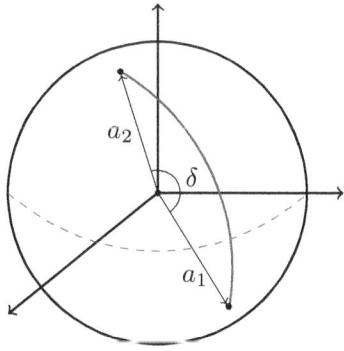

Figure 19.4

$$\cos \delta = \frac{<a_1, a_2>}{|a_1| \cdot |a_2|} = -0.8958 \qquad \Rightarrow \qquad \delta = 2.6811 \text{ radians}$$

Length of the geodesic Valdivia-Yokohama $= 2.6811 \cdot 6367$ Km $= 17071$ Km

As it can be seen, the length following the geodesic (which is NOT sailing with a fixed course) is about 176 km less than following the loxodrome (sailing with a fixed course all the way).

FINAL REMARKS

- If we look at the trajectories of the loxodrome and the geodesic on Mercator's map, a curiosity occurs: the geodesic would not be a straight line joining both points (since this is the loxodrome) but a curve. This is not a contradiction since Mercator's map preserves angles but not distances.

- You have to be careful if you want to repeat the calculation of the previous example for any other two Earth points, since in some cases the shortest loxodrome would not be the line joining both points on Mercator's map. A point with very large longitude (for example, 320°) and another point with very small longitude (for example, 10°) are very far away in Mercator's map, since the Greenwich meridian (in the Pacific Ocean) is the outer border of the map. In those cases we can consider that longitudes are, for example, 30° and 80°, keeping the original difference of 50° but eliminating this problem.

Chapter 20

Kepler's equation

(Kepler – 1609)

PROBLEM

Assuming that we know the position of a planet at a given moment, to calculate at which point of its orbit it will be at any future moment.

HISTORY

Johannes KEPLER (1571 – 1630) was one of the greatest astronomers of all time. This famous problem can be found in chapter 60 of his masterpiece "*Astronomia nova*", published in Prague in 1609.

Statue of Kepler and Brahe (Prague)
Photo: Øyvind Holmstad (Wikimedia Commons)

Kepler was the discoverer of the three laws of planetary orbits that bear his name and since then he looked for a way to calculate the position of the planets assuming a known starting point and a time elapsed from that initial point. If the orbits of the planets had been circular and their angular velocities constant, the problem would be very simple, but Kepler's laws precisely affirm that the orbits are elliptical and the speeds of the planets are different at each moment. Kepler was able to established his famous laws thanks to the precise data of the position of Mars in the sky that the danish astronomer Tycho BRAHE (1546-1601) had collected every night for many years in Prague.

To solve the problem, Kepler defined the terms of average, eccentric and real anomaly; they are only names of concepts that will help to find the solution, as we will see below.

117

Definitions

Let S and P be two points representing the Sun and the planet describing an elliptical orbit around it (the Sun is at one of the foci of the ellipse, as Kepler's laws state). Let N be the point of the orbit of the planet that is closest to the Sun, called **perihelion**; let O be the origin of coordinates and the center of the ellipse and its circumscribed circle (see figure 20.1); let P' be the point of intersection of the circumscribed circle with the parallel that passes through P (coordinates (x, y) to the minor axis of the orbit; let a and b be the lengths of major and minor axis of the ellipse, respectively; let $\overline{OS} = e$ be the so-called linear eccentricity, $\epsilon = e/a$ the astronomical eccentricity, T the period of revolution of the planet and t the elapsed time to reach the position of the planet P since pass through perihelion.

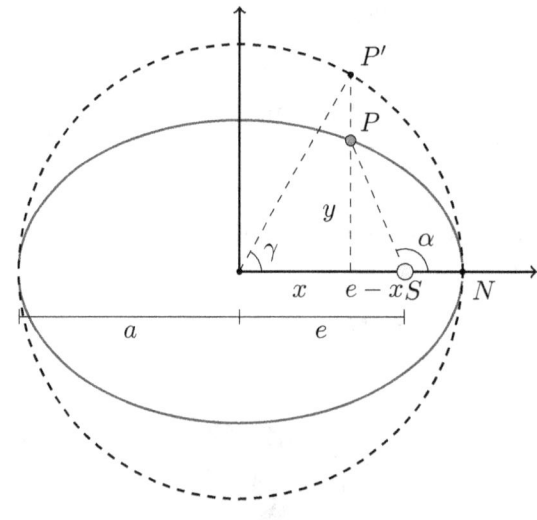

Figure 20.1

The true anomaly α is the angle $\widehat{NSP}$, i.e., the angle described by the focal radius of the planet at time t. This is **the angle we want to calculate** at any time in the future, since it will allow us to know where the planet is.

The average anomaly β is the angle that the focal radius would have described in time t if it grew uniformly (and completing the orbit in the same period of revolution T), so that its value in radians is:

(20.1)
$$\beta = \frac{2\pi}{T} \cdot t$$

This is **the angle we know**, since it is calculated with the time elapsed from a known position on the planet (i.e., variables t - elapsed time - and T -revolution period of the planet - are known). It is an imaginary angle, and that is why it is not represented in figure 20.1.

Finally, the eccentric anomaly γ is the angle $\widehat{NOP'}$ between the radius of the circumscription circle at P' and the radius $\overline{ON}$.

Relation between true and eccentric anomalies

Let us first find the relationship that exists between true anomaly α and eccentric anomaly γ, which, as we have explained, has no real practical meaning (it is only a useful intermediate variable).

Taking γ as a variable, we have that the equation of the orbit can be parameterized as:

$$\begin{cases} x = a\cos\gamma \\ y = b\sin\gamma \end{cases}$$

while the equation of the circumscribed circle is written as:

$$\begin{cases} x = a\cos\gamma \\ y = a\sin\gamma \end{cases}$$

If we look at the right triangle of sides $e - x$ and y of figure 20.1 and apply the above equations:

$$\tan\alpha = \frac{b\sin\gamma}{a\cos\gamma - e} \qquad \Rightarrow \qquad \tan^2\alpha = \frac{b^2\sin^2\gamma}{a^2\cos^2\gamma - 2ae\cos\gamma + e^2} \qquad \Rightarrow$$

$$\Rightarrow \qquad -1 + \sec^2\alpha = \frac{(a^2 - e^2)\sin^2\gamma}{a^2\cos^2\gamma - 2ae\cos\gamma + e^2} = \frac{(1 - \epsilon^2)\sin^2\gamma}{\cos^2\gamma - 2\epsilon\cos\gamma + \epsilon^2} \qquad \Rightarrow$$

$$\Rightarrow \qquad \sec^2\alpha = \frac{\cos^2\gamma - 2\epsilon\cos\gamma + \epsilon^2 + \sin^2\gamma - \epsilon^2\sin^2\gamma}{\cos^2\gamma - 2\epsilon\cos\gamma + \epsilon^2} = \left(\frac{1 - \epsilon\cos\gamma}{\cos\gamma - \epsilon}\right)^2 \qquad \Rightarrow$$

$$\Rightarrow \qquad \cos\alpha = \frac{\cos\gamma - \epsilon}{1 - \epsilon\cos\gamma}$$

For the last step we have been careful in choosing the correct sign after extracting the root and it is easy to verify that this equation is correct for any angle α between 0 and 360°.

From the last equation we can deduce the following two equations:

$$1 - \cos\alpha = \frac{1 - \epsilon\cos\gamma - \cos\gamma + \epsilon}{1 - \epsilon\cos\gamma} = \frac{(1 + \epsilon)(1 - \cos\gamma)}{1 - \epsilon\cos\gamma}$$

$$1 + \cos\alpha = \frac{1 - \epsilon\cos\gamma + \cos\gamma - \epsilon}{1 - \epsilon\cos\gamma} = \frac{(1 - \epsilon)(1 + \cos\gamma)}{1 - \epsilon\cos\gamma}$$

Dividing both equations:

$$\frac{1 - \cos\alpha}{1 + \cos\alpha} = \frac{1 + \epsilon}{1 - \epsilon} \cdot \frac{1 - \cos\gamma}{1 + \cos\gamma}$$

Finally, with the help of the well-known trigonometric formulas of half angles:

$$1 - \cos\phi = 2\sin^2(\phi/2) \qquad \text{and} \qquad 1 + \cos\phi = 2\cos^2(\phi/2)$$

119

we obtain this formula (known as Gauss' formula):

$$\tan\frac{\alpha}{2} = \sqrt{\frac{1+\epsilon}{1-\epsilon}} \cdot \tan\frac{\gamma}{2}$$

Relation between median and eccentric anomalies

Now we look for the relation between β and γ, which will allow us, at the end of the section, to relate α and β, our final goal. The example that we will expose later will help to understand the whole process.

The relationship we are going to find is based on one of Kepler's well-known laws about the orbits of the planets, which he found by observation hundreds of years ago: "*The focal radius of a planet sweeps equal areas in equal times*"

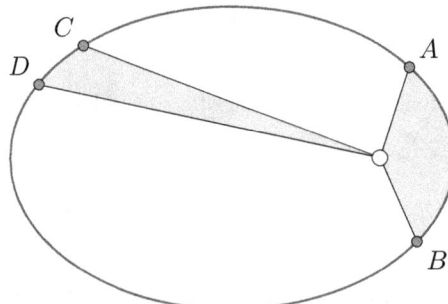

Figure 20.2

In figure 20.2 we see the idea that Kepler wanted to convey: when the planet is far from the Sun, the focal radius covers a lot of surface when it moves a few degrees; furthermore, the closer it is to it, the greater the angle needed to achieve the same surface. Kepler's law implies that if both areas are equal, the planet takes the same time to go from A to B that from C to D, although in the latter case the distance traveled by the planet is much smaller.

That is, the planet goes faster in its orbit when it is closer to the Sun, since it is obliged to comply with Kepler's law and therefore it has to move faster to cover the same surface as when it was far from the Sun. This law is inferred (though it is not trivial) from Newton's formula of universal attraction.

From the law it is inferred that there is a direct proportion between the area and the time taken by the planet to sweep: double area needs double amount of time, triple area needs triple amount of time, etc. We are going to apply this proportion for two determined areas: the area of the orbit (in figure 20.1) enclosed by points SNP (let J be the value of this area and let t the time that the planet takes to traverse it) and the area of the entire orbit (which has the value of the area of an ellipse, i.e., the value πab, and the planet takes a value of time T - its period of revolution - to traverse it). Then:

(20.2)
$$\frac{J}{t} = \frac{\pi ab}{T} \qquad \Rightarrow \qquad J = \frac{ab}{2}\cdot\left(2\pi\cdot\frac{t}{T}\right) = \frac{ab}{2}\cdot\beta$$

where we have applied equation (20.1). We only need to find the value of the area J as a function of the eccentric anomaly γ in order to find our sought relation when we replace it in equation (20.2).

Let us look at figure 20.3 and try to figure out how to calculate areas J_1, J_2 and J.

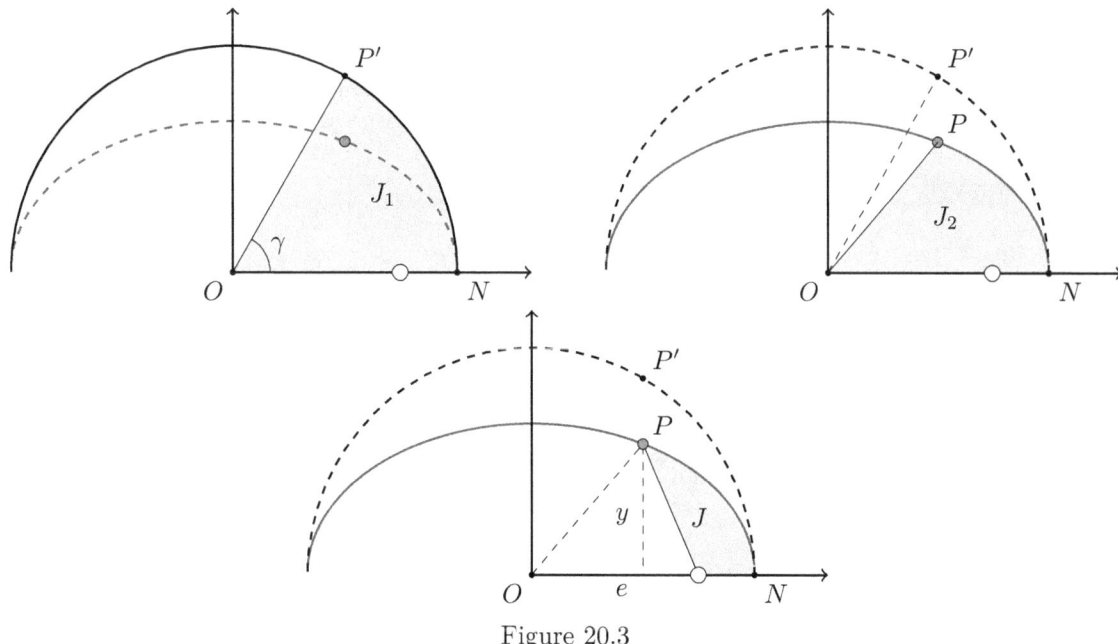

Figure 20.3

First, the area J_1 can be calculated with a simple ratio: if, for an angle of 2π, we have that the area of the circle of radius a is equal to $\pi \cdot a^2$, then, for an angle γ, we have that the area is equal to $(\pi \cdot a^2) \cdot (\gamma/2\pi)$. That means that $J_1 = (\pi \cdot a^2) \cdot (\gamma/2\pi) = a^2\gamma/2$.

Now, for the area J_2 we have to think in the following way: the area J_2 is the result of a proportional and vertical "flattening" of the area J_1: if a segment that is parallel to the axis y has a length of a, it would have a length of b after the operation (this property is clearly seen in the axis y, where the radius of the circumference a is flattened until becoming the semi-axis of the ellipse of length b; the rest of the ordinates of points of the circumference and the ellipse - for example, ordinate of point P' is flattened to become ordinate of point P - follow the same proportion). Therefore, that means that the value of the area J_2 is equal to the value of the area J_1 multiplied by b/a, since a square of side equal to 1 in J_1, oriented according to the axes, is transformed into a rectangle of sides 1 (the side that is parallel to the x axis, which does not change) and b/a (the side that is parallel to the y axis, flattened as explained above). That is, $J_2 = J_1 \cdot (b/a) = (ab\gamma)/2$.

Mathematically we get the same conclusion if we let $f(x)$ be the function that surrounds area J_1 determined by O, P' and N (therefore, $f(x)$ consists of a segment attached to a piece of circumference) and let $g(x)$ be the function that surrounds area J_2 determined by O, P and N ($g(x)$ is a segment attached to a piece of ellipse). Since $f(x) = (a/b) \cdot g(x)$, it is easy to see that:

$$J_2 = \int_0^N g(x)dx = \int_0^N \frac{b}{a} \cdot f(x)dx = \frac{b}{a} \int_0^N f(x)dx = \frac{b}{a} \cdot J_1 = \frac{b}{a} \cdot \frac{1}{2}a^2\gamma = \frac{ab}{2} \cdot \gamma$$

Finally, the area J is equal to the area J_2 subtracting the area defined by triangle OPS:

$$J = J_2 - \frac{ey}{2} = \frac{ab\gamma}{2} - \frac{a\epsilon b \sin\gamma}{2} = \frac{ab}{2} \cdot (\gamma - \epsilon \sin\gamma)$$

Replacing this expression in (20.2), we finally obtain Kepler's equation:

(20.3) $$\gamma - \epsilon \sin\gamma = \beta$$

Approximate solution to Kepler's equation

Suppose we know the value of β (i.e., we know the time elapsed since perihelion) and the eccentricity of the planet's orbit ϵ, and we want to calculate the value of γ that satisfies Kepler's equation. The solution cannot be calculated exactly (there is no way to isolate γ from equation (20.3)), so an iterative method of calculation is proposed.

Let us calculate the values γ_1, γ_2, γ_3, ... so that we will quickly approximate to the value of γ, making the error negligible. We define the values of γ_1, γ_2, γ_3, ... with the following iterations:

$$\gamma_1 = \beta + \epsilon \sin \beta$$
$$\gamma_2 = \beta + \epsilon \sin \gamma_1$$
$$\gamma_3 = \beta + \epsilon \sin \gamma_2$$
$$...$$

The error committed in the first value γ_1 with respect to the sought value can be bounded as follows:

$$|\gamma - \gamma_1| = |\epsilon(\sin \gamma - \sin \beta)| \overset{(1)}{\leq} \epsilon \cdot |\gamma - \beta| = \epsilon^2 \cdot |\sin \gamma| \leq \epsilon^2$$

To explain inequality (1) we must apply the theorem of the mean value for derivable functions: if we have a derivable function $f(x)$ in an open interval it is satisfied that, for all $a < b$ points within the interval, there exists a value c (where $a < c < b$) such that $[f(a) - f(b)]/[ab] = f'(c)$. If we apply it to function $f(x) = \sin x$ in the interval $(0, 2\pi)$, then we have that $(\sin \gamma - \sin \beta)/(\gamma - \beta) = f'(c)$ which, taking absolute values in both sides, leads us to $|\sin \gamma - \sin \beta| = |f'(c)| \cdot |\gamma - \beta|$. Now we just keep in mind that $|f'(c)| \leq 1$ (since the derivative of $f(x) = \sin x$ is $f'(x) = \cos x$ and this function is bounded by 1) to deduce the inequality.

The error committed in the second value can be bounded in a similar way:

$$|\gamma - \gamma_2| = |\epsilon(\sin \gamma - \sin \gamma_1)| \leq \epsilon \cdot |\gamma - \gamma_1| \leq \epsilon^3$$

Following an identical reasoning we find the maximum errors of each iteration, and they approximate to 0 since $\epsilon < 1$. In Earth's case, its orbit has an eccentricity of $\epsilon = 0.01674$ resulting in $\epsilon^3 = 0.00000469$. Taking into account that one second of degree is equivalent, in radians, to 0.00000485, the approximation of γ_3 is exact up to one second of degree.

As a final conclusion, Kepler's and Gauss' formulas allow to obtain the values of anomalies γ and α from the mean anomaly β, i.e., from the time that has elapsed since perihelion, with equation (20.1).

Example

"*Assuming Earth is in the perihelion of its orbit on January 1, 2008 at 0:00 GMT, look for its position in the orbit by April 23, 2008 at 19:17 GMT*"

A sidereal year (the amount of time it takes to Earth to make a round to its orbit) lasts $365,256$ 24-hour days (hence the need for a leap year every four years). 2008 was a leap year, so at the time indicated, $113,803$ 24-hour days had elapsed.

Therefore, the mean anomaly β at that time was equal to:

$$\beta = \frac{113.803}{365.256} \cdot 2\pi = 1.958 \qquad (= 112.19°)$$

To calculate the eccentric anomaly, we use the iterative method of the previous section:

$$\gamma_1 = \beta + \epsilon \sin \beta = 1.958 + 0.01674 \cdot \sin 1.958 = 1.973$$
$$\gamma_2 = \beta + \epsilon \sin \gamma_1 = 1.958 + 0.01674 \cdot \sin 1.973 = 1.973$$

More iterations are no longer necessary because we have found the same value for γ_1 and γ_2 and therefore $\gamma = 1.973$ ($= 113.07°$). Finally, to calculate the real anomaly:

$$\tan \frac{\alpha}{2} = \sqrt{\frac{1+\epsilon}{1-\epsilon}} \cdot \tan \frac{\gamma}{2} = \sqrt{\frac{1+0.01674}{1-0.01674}} \cdot \tan \frac{1.973}{2} \qquad \Rightarrow$$

$$\Rightarrow \quad \alpha = 2\arctan(1.538) = 1.988 \qquad (= 113.95°)$$

At the indicated day and time, Earth is at the point of the elliptical orbit such that the angle between Earth, the Sun and the perihelion is $113.95°$.

Because the eccentricity of Earth is very small (i.e., its orbit is almost circular), there is little difference among the three anomalies. In other orbits (for example, Mars) the difference would be much greater.

FINAL REMARKS

- Kepler's equation allows us to calculate the position of Earth relative to the Sun at any time in advance, i.e. the so-called equation of time e that is reflected in some maritime tables collected in "almanacs". Among other applications, the equation of time calculated in these tables was necessary to know the position of a ship, as we will see in the problem "Calculation of the position at sea".

- Kepler's laws were deduced from the observation of data and they fitted with real data thanks to a correct hypothesis. For example, when looking for the first law, at some point Kepler assumed that the orbits were elliptical and that the Sun was in a focus (perhaps after realizing that other simpler hypotheses - as circular orbits or elliptical orbits with the Sun at the center of the ellipse - did not match the data) and found that Brahe's observations matched perfectly with that assumption. But it seems that he did not understand the reason why the laws were satisfied.

- It had to be the great English scientist Isaac Newton (1643 – 1727), for many the most brilliant genius of history, who would mathematically deduce Kepler's laws from the Law of Universal Gravitation (1684) that he had recently discover. One of the great questions in the history of science was thus discovered.

Chapter 21

Expanding a map

(Snellius – 1617)

PROBLEM

To determine the position of new points (discovered by explorers) on a map, using the positions of other known points.

HISTORY

This problem was of great importance several centuries ago, as that time it was important to add new points on precise maps.

For this problem, we define an "accessible point" as a point which a explorer can easily reach and calculate its distance to other accessible points, and an "inaccessible point" as a point which a explorer cannot reach (for example, because there is water in between). When studying an inaccessible point the explorer does not know the distance to it, but he can see it from where he is and he can calculate the angle between a fixed direction and the direction toward it.

Willebrord Snel
Portrait as University teacher

Terrestrial and marine explorers frequently encountered the following two cases:

- Problem of Snellius–Pothenot or problem of the three inaccessible points: *To determine the position of an unknown accessible point on a map by using the positions of three known points A, B and C which are now all inaccessible points.*

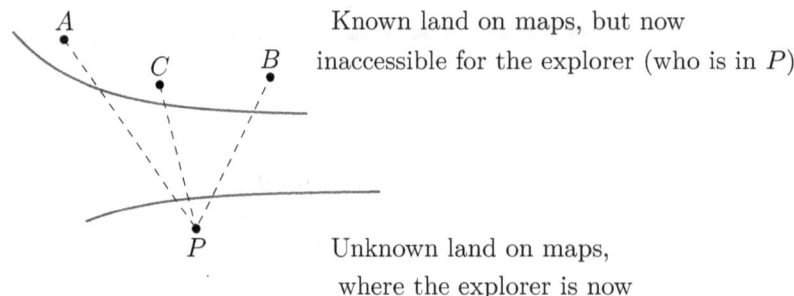

Figure 21.1

This problem, the most famous one among all problems of terrestrial exploration, was stated and solved by Dutchman Willebrord Snel (1581 – 1626, also known as Snellius) in his work "Eratosthenes Batavus" (1617), but it did not attract the attention of his contemporaries. It was not widely known until it was solved again by Frenchman Pothenot (who died in 1732) in a paper sent to the French Academy in 1692. It has been known as Photenot's problem since then.

- Hansen's problem or problem of the inaccessible distance: *From the position of two inaccessible known points A and B, to determine the position of two unknown points P and P'.*

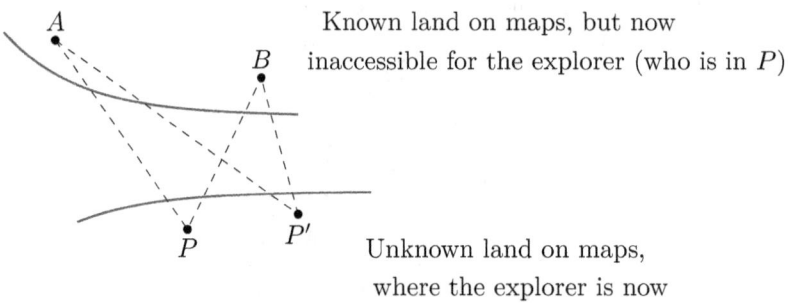

Figure 21.2

Compared to the previous case, here there are only two good reference points (lighthouses, houses, etc.) that could be seen from accessible land but they are not yet on known maps. Some people thought of this variant of the problem, solved by German astronomer Hansen (1795 – 1874) but also by other authors before him.

SOLUTION

First we will deduce a trigonometric formula that will be useful later:

LEMMA 21.1. *Let α and β be two angles that satisfy the equation $\sin\alpha/\sin\beta = m/n$. Then:*

$$\frac{\tan\frac{\alpha-\beta}{2}}{\tan\frac{\alpha+\beta}{2}} = \frac{m-n}{m+n}$$

PROOF. Let us first prove the following equality:

(21.1)
$$\frac{\sin \alpha - \sin \beta}{\sin \alpha + \sin \beta} = \frac{m - n}{m + n}$$

We can do so using the following reasoning:

$$\frac{\sin \alpha - \sin \beta}{\sin \alpha + \sin \beta} = \frac{\frac{\sin \alpha - \sin \beta}{\sin \beta}}{\frac{\sin \alpha + \sin \beta}{\sin \beta}} = \frac{\frac{\sin \alpha}{\sin \beta} - 1}{\frac{\sin \alpha}{\sin \beta} + 1} \overset{(a)}{=} \frac{\frac{m}{n} - 1}{\frac{m}{n} + 1} = \frac{m - n}{m + n}$$

where we have applied in (a) the hypothesis of the lemma.

We now take advantage of the well-known trigonometric formula of addition and subtraction of sines, whose proof can be easily found by the reader:

(21.2)
$$\sin \alpha \pm \sin \beta = 2 \cdot \sin \left(\frac{\alpha \pm \beta}{2} \right) \cdot \cos \left(\frac{\alpha \mp \beta}{2} \right)$$

If we replace in equation (21.1) what we have found in equation (21.2), then:

$$\frac{\sin \alpha - \sin \beta}{\sin \alpha + \sin \beta} = \frac{m - n}{m + n} \quad \Rightarrow \quad \frac{2 \cdot \sin \left(\frac{\alpha - \beta}{2} \right) \cdot \cos \left(\frac{\alpha + \beta}{2} \right)}{2 \cdot \sin \left(\frac{\alpha + \beta}{2} \right) \cdot \cos \left(\frac{\alpha - \beta}{2} \right)} = \frac{m - n}{m + n} \quad \Rightarrow$$

$$\Rightarrow \quad \frac{\tan \frac{\alpha - \beta}{2}}{\tan \frac{\alpha + \beta}{2}} = \frac{m - n}{m + n}$$

$\square$

Pothenot's problem

In this problem we assume that we know the distances between known points AC and BC (because they are included on known maps), and we give them the values of a and b, respectively. We can assume that we know the angle $\overparen{ACB}$ too, and let us say that its value is γ.

Now, we (the explorers) are at point P (unknown on maps) so we are able to calculate the angle $\overparen{APC}$ (α) and the angle $\overparen{BPC}$ (β). By hypothesis, it is assumed that points A, B and C are inaccessible at this time, which means that the distances $\overline{AP}$ (x), $\overline{BP}$ (y) and $\overline{CP}$ (z) are unknown, as well as the angles $\overparen{CAP}$ (ψ) and $\overparen{CBP}$ (ϕ).

In figure 21.3 all the variables are represented (some of them are known, others are unknown). Our goal is to determine these unknown variables since once they are determined we will add the point P to our maps with total accuracy.

If we apply the sine theorem to the triangles ACP and BCP we deduce the following equations:

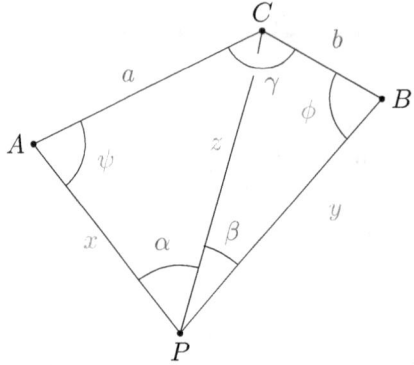

Figure 21.3

$$\frac{\sin \psi}{\sin \alpha} = \frac{z}{a} \quad \text{and} \quad \frac{\sin \phi}{\sin \beta} = \frac{z}{b}$$

Dividing both of them:

$$\frac{\sin \psi}{\sin \phi} = \frac{b \cdot \sin \alpha}{a \cdot \sin \beta}$$

The right-hand side of the equality is a known value, so let us denote it by μ for simplicity:

$$\frac{\sin \psi}{\sin \phi} = \mu$$

We now apply lemma 21.1 with $m = \mu$ and $n = 1$:

(21.3)
$$\frac{\tan \frac{\psi - \phi}{2}}{\tan \frac{\psi + \phi}{2}} = \frac{\mu - 1}{\mu + 1}$$

Since the value of the angle $\psi + \phi$ is known (it is equal to $360° - \alpha - \beta - \gamma$, all of them known values), the only unknown variable in equation (21.3) is the value of $(\psi - \phi)/2$, so we can isolate it and calculate it. After that it is easy to calculate the values of ψ and ϕ separately (using the information we know about their addition and subtraction).

Finally, the unknown values x, y, z can be obtained from the following formulas (applying the sine theorem):

$$\frac{x}{a} = \frac{\sin(180° - \alpha - \psi)}{\sin \alpha} \qquad \frac{y}{b} = \frac{\sin(180° - \beta - \phi)}{\sin \beta} \qquad \frac{z}{a} = \frac{\sin \psi}{\sin \alpha}$$

The position of P can be now perfectly drawn on the map thanks to the magnitudes ψ, ϕ, x, y, and z.

Hansen's problem

In this problem we assume that we know the distance $\overline{AB}$, which we call c, because A and B are known points on the map. Furthermore, we (the explorers) can reach both unknown points P and P' so we can calculate the angles $\widehat{APB}$, $\widehat{AP'B}$, $\widehat{BPP'}$ and $\widehat{AP'P}$ (we call them γ, γ', δ and δ', respectively).

It is assumed by hypothesis that points A and B are now inaccessible, which means that the distances $\overline{AP}$ (x), $\overline{AP'}$ (x'), $\overline{BP}$ (y) and $\overline{BP'}$ (y') are unknown, as well as the values of the angles $\widehat{BAP'}$, $\widehat{ABP}$ (ψ and ϕ, respectively).

The angles $\widehat{PAP'}$ and $\widehat{PBP'}$ (α and β, respectively) are known, since each one completes a triangle where the values of the other two angles are known. Finally, we can assume that the distance $\overline{PP'}$, which we call s, is unknown (the problem can be solved without calculating it).

In figure 21.4 all variables are represented (some of them are known, others are unknown). Our goal is to determine these unknown variables since once they are determined we will add the point P to our map with total accuracy.

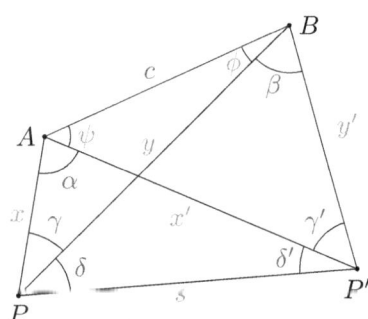

Figure 21.4

Applying the sine theorem to triangles BAP, APP', $PP'B$ and $P'BA$:

$$\frac{\sin\gamma}{\sin\phi} = \frac{c}{x} \qquad \frac{\sin\delta'}{\sin\alpha} = \frac{x}{s} \qquad \frac{\sin\beta}{\sin\delta} = \frac{s}{y'} \qquad \frac{\sin\psi}{\sin\gamma'} = \frac{y'}{c}$$

Now multiplying all three equations:

$$\frac{\sin\psi \cdot \sin\beta \cdot \sin\gamma \cdot \sin\delta'}{\sin\phi \cdot \sin\alpha \cdot \sin\gamma' \cdot \sin\delta} = 1 \qquad \Rightarrow \qquad \frac{\sin\psi}{\sin\phi} = \frac{\sin\alpha \cdot \sin\gamma' \cdot \sin\delta}{\sin\beta \cdot \sin\gamma \cdot \sin\delta'}$$

The right-hand side of the previous equality is a known value, so let us denote it by μ for simplicity:

$$\frac{\sin\psi}{\sin\phi} = \mu$$

We now apply lemma 21.1 with $m = \mu$ and $n = 1$:

$$(21.4) \qquad \frac{\tan \frac{\psi - \phi}{2}}{\tan \frac{\psi + \phi}{2}} = \frac{\mu - 1}{\mu + 1}$$

Let D be the point where the diagonals of the quadrilateral $ABP'P$ meet. Then, the angles $\widehat{ADB}$ and $\widehat{PDP'}$ are equal; that means that the value $\psi + \phi = \delta + \delta'$, i.e., $\psi + \phi$ is known. Therefore, the only unknown value in equation (21.4) is $(\psi - \phi)/2$, so we isolate it and calculate it. After that it is easy to calculate the values of ψ and ϕ separately (using the information we know about their addition and subtraction).

Once the values of ψ and ϕ are known, it is easy to calculate the rest of the unknown angles (looking at the triangle ABP, it follows that $\gamma + \phi + \alpha + \psi = 180$ and from there we find the value of α; in a similar way, we use the triangle ABP' to find the value of β).

Finally, the unknown values x, y, x', y' and s are obtained from the following formulas (using the sine theorem):

$$\frac{\sin \gamma}{\sin \phi} = \frac{c}{x} \qquad \frac{\sin \delta'}{\sin \alpha} = \frac{x}{s} \qquad \frac{\sin \beta}{\sin \delta} = \frac{s}{y'} \qquad \frac{\sin \alpha}{\sin(\delta + \phi)} = \frac{s}{x'} \qquad \frac{\sin \beta}{\sin(\delta' + \gamma')} = \frac{s}{y}$$

The position of P can be now perfectly drawn on the map thanks to the magnitudes ψ, ϕ, β, α, x and y, while the position of P' can be drawn too thanks to the magnitudes ψ, ϕ, β, x' and y'.

FINAL REMARKS

Mathematical knowledge was an important part of terrestrial and maritime expeditions at Photelot and Hansen's times, also considering that large tables were necessary to calculate trigonometric functions. After studying this solution, we can imagine antique expeditionary spending many hours with laborious calculations, trying to improve the maps of that era.

Chapter 22

Area of a section of a hyperbola

(Saint-Vicent – 1630)

PROBLEM

To calculate the area of a section of a hyperbola.

HISTORY

In the problem "Area of a section of a parabola" we discussed how Archimedes found (in 240 B.C.) a wonderful method to calculate the area enclosed within a section of a parabola. To achieve the same result with the hyperbola, one of the conic sections, we had to wait many centuries.

The first man that began to sketch the solution was a Jesuit monk named Grégoire de Saint-Vicent (1584 – 1667), born in the Belgian city of Bruges, at that time part of the Spanish Empire.

Grégorie de Saint-Vicent
Portrait of the Jesuit order

In his study of how to calculate areas, Saint-Vicent used a method of approximation that we can now compare to the modern integral calculus, which was developed by Leibniz and Newton much later. So we consider him to be a man ahead of his time.

However, the solution to the problem that we are going to study is one made after Saint-Vicent, although it does not use integral calculus. Its difficulty and beauty gives us an idea of the imagination and technique that mathematicians had then.

SOLUTION

Appropriate coordinate system

Let us set a hyperbola whose major axis coincides with the X axis, whose minor axis coincides with the Y axis, and let a and b be the values of the lengths of its semi-major and semi-minor axes, respectively. In this case, it is well known that the equation of the hyperbola can be written as:

(22.1)
$$\frac{x^2}{a^2} - \frac{y^2}{b^2} = 1$$

Let α be the angle between the X axis and any of the asymptotes of the hyperbola (i.e., $\tan\alpha = b/a$) and let c be the value of $\sqrt{(a^2 + b^2)}$, so $\cos\alpha = a/c$ and $\sin\alpha = b/c$.

We want to find the area A of the section of the hyperbola enclosed by the line that is at distance d from the origin of coordinates $(d > a)$. Let H and K be the vertices of the section, whose coordinates are (d, f) and $(d, -f)$ respectively, as we can see in figure 22.1.

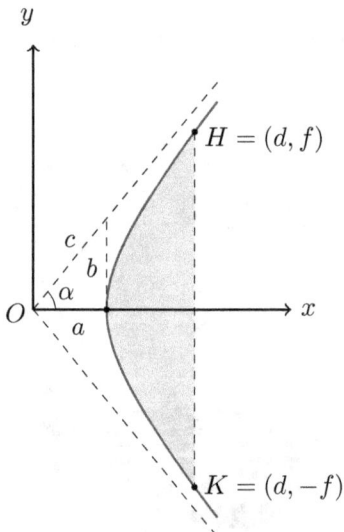

Figure 22.1

It should be noted that if we can find a formula to calculate this area, any other area enclosed by the hyperbola and two of its points could be found by applying this formula and a reasoning similar to that used in the problem of the Archimedes' parabola.

Saint-Vicent realized that he had to first look for a more suitable coordinate system, where the asymptotes of the hyperbola (drawn in figure 22.1 with dashed lines, with equations $y = \pm(b/a)\cdot x$) are the new coordinate axes, while maintaining the same point O as the origin of coordinates. This new coordinate system is not orthogonal (that is, its axes are not perpendicular between them), so we will have to be careful in its treatment.

Once the new coordinate axes have been chosen, the scale to be used must be decided: Saint-Vicent transformed the points (a, b) and $(a, -b)$, in the old coordinate system (x, y), into the points $(0, c)$ and $(c, 0)$ in the new coordinate system (u, v). This choice is the most appropriate one, since we

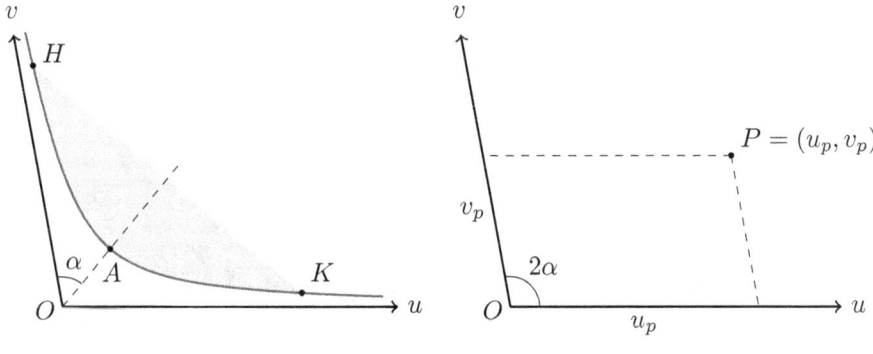

Figure 22.2

have transformed the vector (a, b) into the vector $(0, c)$ (both of equal length c) and the vector $(a, -b)$ into the vector $(c, 0)$ (both of equal length c too). By keeping the length of these two vectors (that can be considered as a basis of the coordinate system in each system), any length of any other vector will also remain identical and, as a consequence, the area of any figure remains constant in both coordinate systems.

The reader can verify that the linear application that transforms the coordinates of system (x, y) into the coordinates in system (u, v) is:

(22.2)
$$\begin{cases} u = (c/2a) \cdot x + (-c/2b) \cdot y \\ v = (c/2a) \cdot x + (c/2b) \cdot y \end{cases}$$

and its inverse function is:

(22.3)
$$\begin{cases} x = (a/c) \cdot u + (a/c) \cdot v \\ y = (-b/c) \cdot u + (b/c) \cdot v \end{cases}$$

For readers that are not familiar with non-orthogonal coordinate systems, the coordinates of any point $P = (u_p, v_p)$ are calculated graphically by drawing lines from that point that are parallel to the coordinate axes: distances from the origin O to the points of intersection of these lines are the coordinates u_p and v_p (see the right graph of figure 22.2).

It should also be noted that, for the calculation of areas of parallelograms in the new coordinate system, the angle between both coordinate axes has a value of 2α; for example, the area of the parallelogram on the right side of figure 22.2 is equal to $u_p \cdot v_p \cdot \sin(2\alpha)$.

But, why is this coordinate system suitable for solving the problem? The reason is that, in this coordinate system, the hyperbola equation is very simple: if we replace equations (22.3) in formula (22.1), we find out that the new hyperbola equation, now in the coordinates (u, v), becomes:

(22.4)
$$\frac{[(a/c) \cdot u + (a/c) \cdot v]^2}{a^2} - \frac{[(-b/c) \cdot u + (b/c) \cdot v]^2}{b^2} = 1 \quad \Rightarrow \quad \cdots \quad \Rightarrow \quad u \cdot v = \frac{c^2}{4}$$

That is, we can forget about equation (22.1) and its annoying quadratic terms, and move on to a simpler equation such as (22.4), which will allow us to easily calculate the sought area.

133

In the new coordinate system, the points A, H and K have the following coordinates (applying equation 22.2):

$$(22.5) \qquad A = \frac{c}{2} \cdot (1,1) \qquad H = \frac{c}{2} \cdot \left(\frac{d}{a} - \frac{f}{b}, \frac{d}{a} + \frac{f}{b} \right) \qquad K = \frac{c}{2} \cdot \left(\frac{d}{a} + \frac{f}{b}, \frac{d}{a} - \frac{f}{b} \right)$$

The values d and f are related to each other, since the point (d, f) lies in the hyperbola and, therefore, its coordinates satisfy equation (22.1). As this equation can also be written as:

$$\left(\frac{x}{a} - \frac{y}{b} \right) \cdot \left(\frac{x}{a} + \frac{y}{b} \right) = 1$$

that means that the coordinates (d, f) satisfy the equation:

$$(22.6) \qquad \left(\frac{d}{a} - \frac{f}{b} \right) \cdot \left(\frac{d}{a} + \frac{f}{b} \right) = 1$$

which implies that the multiplication of coordinates of point H (and point K too) results in the value $c^2/4$, as it should be to satisfy equation (22.4) because those points are points of the hyperbola (point A also clearly satisfy the same equation, of course).

Relations between areas

The next step of Saint-Vicent's method is how to calculate the area sought in the new coordinate system. But before that he studied the different areas involved in the problem. Let us look at figure 22.3.

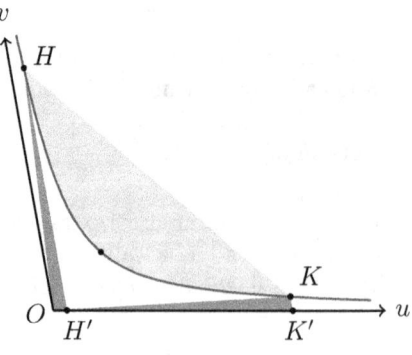

Figure 22.3

In figure (22.3) we have two triangles: the first one has vertices at points O, H and at the intersection of axis u with the line that is parallel to axis v and passes through H (let H' be this intersection), while the second one has vertices at points O, K and at the intersection of axis u with the line that is parallel to axis v and passes through K (let K' be this intersection).

The area of the triangle OHH' can be calculated (using the second graph of figure 22.2) as half of the multiplication of the coordinates of point H and the value $\sin(2\alpha)$, i.e., (applying 22.5 and 22.6):

$$\sin(2\alpha) \cdot u_h \cdot v_h = \sin(2\alpha) \cdot \frac{c}{2} \cdot \left(\frac{d}{a} - \frac{f}{b}\right) \cdot \frac{c}{2} \left(\frac{d}{a} + \frac{f}{b}\right) = sin(2\alpha) \cdot \frac{c^2}{4}$$

Similarly, the area of triangle OKK' can be calculated as half of the multiplication of the coordinates of point K and the value $\sin(2\alpha)$:

$$\sin(2\alpha) \cdot u_k \cdot v_k = \sin(2\alpha) \cdot \frac{c}{2} \cdot \left(\frac{d}{a} + \frac{f}{b}\right) \cdot \frac{c}{2} \left(\frac{d}{a} - \frac{f}{b}\right) = sin(2\alpha) \cdot \frac{c^2}{4}$$

That is, the values of the areas of both triangles are equal.

Now let us look at figure 22.4: the previous result shows us that the area drawn in the first graph must be equal to the area drawn in the second graph, since to get the second area from the first one we have removed the area of triangle OHH' and then we have added the area of triangle OKK'. But we have seen that these two areas are equal, hence the result.

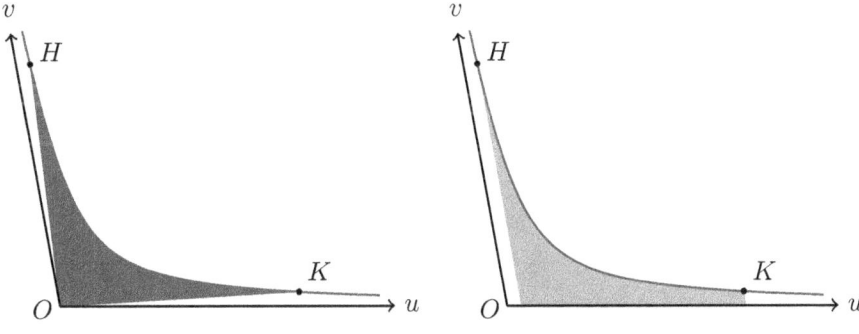

Figure 22.4

So Saint-Vicent came to the conclusion that if he could calculate the area of the second graph of figure 22.4, that would be enough to solve the problem.

Therefore, the next step of the method is to calculate that area, which we call a **hyperbolic trapezoid** (and which today we would calculate with integral calculus).

Calculation of the area under the hyperbola

Suppose that the value of the area of the hyperbolic trapezoid is T. We can divide that area into n different parts, so that each of the areas has a value equal to T/n: we can get it by dividing the region with $n-1$ lines that are parallel to the v axis between points H and K, as shown in figure 22.5. The distance between two consecutive lines is not the same, because they are closer near point H (where the height of the section is larger) and they are more distant near point K (where the height of the section is smaller). It is trivial to see that this division of the area can easily be achieved.

Let (u_0, v_0) be the coordinates of the point H, let (u_1, v_1), (u_2, v_2), ..., (u_{n-1}, v_{n-1}) be the coordinates of the points of the hyperbola where the parallel lines (mentioned in the previous paragraph) intersect with it, and let (u_n, v_n) be the coordinates of the point K.

As the hyperbola in the coordinate system (u, v) can be seen as a decreasing function (see the right figure of figure 22.5, which is an enlarged picture of a part of the left figure), the parallelogram of vertices $(u_i, 0)$, $(u_{i+1}, 0)$, (u_i, v_i) and (u_{i+1}, v_i) has an area whose value is greater than the value of the area of the section between the hyperbola, line $u = u_i$, line $u = u_{i+1}$ and line $v = 0$ (area

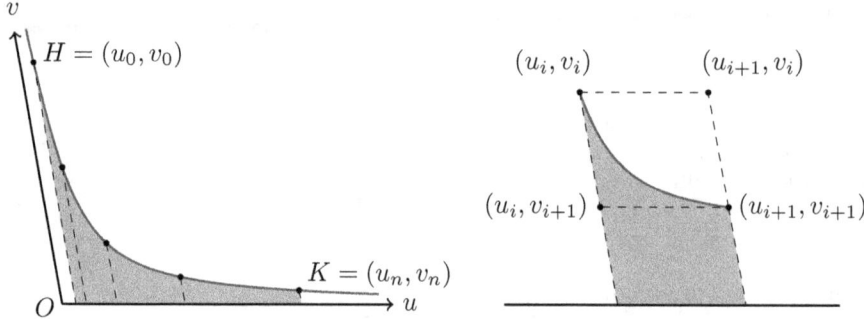

Figure 22.5

in black, its value is equal to T/n), while the parallelogram of vertices $(u_i, 0)$, $(u_{i+1}, 0)$, (u_i, v_{i+1}) and (u_{i+1}, v_{i+1}) has an area whose value is greater than the value of the area of that same section.

Then, we can write that:

$$\text{Bigger parallelogram area} = \sin(2\alpha) \cdot v_i \cdot (u_{i+1} - u_i) > \frac{T}{n}$$

$$\text{Smaller parallelogram area} = \sin(2\alpha) \cdot v_{i+1} \cdot (u_{i+1} - u_i) < \frac{T}{n}$$

Both expressions can be combined in inequalities:

$$\sin(2\alpha) \cdot v_{i+1} \cdot (u_{i+1} - u_i) < \frac{T}{n} < \sin(2\alpha) \cdot v_i \cdot (u_{i+1} - u_i) \qquad \Rightarrow$$

$$\Rightarrow \qquad v_{i+1} \cdot (u_{i+1} - u_i) < \frac{T}{n \cdot \sin(2\alpha)} < v_i \cdot (u_{i+1} - u_i) \qquad \Rightarrow$$

$$\Rightarrow \qquad v_{i+1} \cdot u_{i+1} \cdot \left(1 - \frac{u_i}{u_{i+1}}\right) < \frac{T}{n \cdot \sin(2\alpha)} < v_i \cdot u_i \cdot \left(\frac{u_{i+1}}{u_i} - 1\right)$$

Applying (22.4) in the last inequalities (we can do it because all points belong to the hyperbola) and defining

$$L = \frac{4T}{c^2 \cdot \sin(2\alpha)} \qquad \text{and} \qquad q_i = \frac{u_{i+1}}{u_i}$$

we finally obtain:

$$\left(1 - \frac{1}{q_i}\right) < \frac{L}{n} < (q_i - 1)$$

and, after the isolation of q_i in both inequalities:

(22.7) $$1 + \frac{L}{n} < q_i < \frac{1}{1 + \left(\frac{-L}{n}\right)}$$

Applying this formula for $i = 0, 1, ..., n-1$ we get n inequalities. By multiplying all of them, we obtain:

$$\left(1 + \frac{L}{n}\right)^n < \frac{u_n}{u_0} < \frac{1}{\left(1 + \frac{-L}{n}\right)^n}$$

As we will see in problem "The number e" (in the second part of this book), if we increase the value of n infinitely (i.e., we approximate the parallelograms to the hyperbola more and more), both parts of the inequality have the value of e^L as its limit (e is the Euler number, $e = 2.71828$), so, necessarily, the value of u_n/u_0 has to be equal to that value. That is:

$$\frac{u_n}{u_0} = e^L \quad \Rightarrow \quad L = \ln\left(\frac{u_n}{u_0}\right) \quad \Rightarrow$$

$$\Rightarrow \quad T = \frac{c^2 \cdot \sin(2\alpha)}{4} \cdot \ln\left(\frac{u_n}{u_0}\right) \overset{*}{=} \frac{ab}{2} \cdot \ln\left(\frac{u_n}{u_0}\right)$$

where in (*) we have applied that $\sin(2\alpha) = 2\sin\alpha\cos\alpha = 2\cdot(a/c)\cdot(b/c)$, as it can be deduced from the formula of the double angle and using figure 22.1. Finally, we saw in (22.5) the coordinates of points H and K, from which we deduce that:

$$T = \frac{ab}{2} \cdot \ln\left(\frac{\frac{d}{a} + \frac{f}{b}}{\frac{d}{a} - \frac{f}{b}}\right) \overset{*}{=} \frac{ab}{2} \cdot \ln\left(\left(\frac{d}{a} + \frac{f}{b}\right)^2\right) = ab \cdot \ln\left(\frac{d}{a} + \frac{f}{b}\right)$$

where we have applied formula (22.6) in step (*).

Calculation of the area of the hyperbola

It only remains to complete the method to calculate the target area A (black area in figure 22.1). As already stated above, this value is equal to the value of the area of triangle OHK minus the area calculated in the previous section. Therefore, we have achieved the following remarkable and simple formula for the sought area:

$$A = df - ab \cdot \ln\left(\frac{d}{a} + \frac{f}{b}\right)$$

FINAL REMARKS

Actually, Saint-Vicent did not use logarithms and his method was limited only to square hyperbolas (those whose asymptotes form a right angle), but I wanted to be as faithful as possible to his approach to the problem (the problem has no difficulty if we use integral calculus).

However, we do have to recognize Saint-Vicent as the discoverer of the following property: the value of the area under the hyperbola $x \cdot y = k$ in the interval $[a, b]$ is the same as the value of the area under the hyperbola in the interval $[c, d]$ if the condition $a/b = c/d$ is met. His disciple Alphonse Antonio de Sarasa (1618 – 1667) realized that there is a connection between this property and the logarithmic functions (which were studied independently at that time), as we explained in the problem "Power series of the logarithmic function".

Chapter 23

Torricelli's point

(Torricelli – 1642)

PROBLEM

To find the interior point of a triangle whose sum of distances to the three vertices is a minimum.

HISTORY

According to the legend, this problem was proposed by the French mathematician Pierre Fermat (1601 – 1665) to the Italian physicist Evangelista Torricelli (1608 – 1647), but there is no documentary evidence to support this claim. What we know now is that Torricelli found several solutions to the problem, which were published later by one of his disciples.

Torricelli was a student of Bendetto Castelli, who was in turn a pupil of Galileo; when the latter knew about the works of Torricelli, he invited him to Florence in order to work together (which was obviously a great honor). Unfortunately, Torricelli did not arrive to Florence until 1641 and Galileo died in January 1642, so their collaboration was very short in time.

After the death of Galileo, Torricelli was named his successor as the mathematician of the court of the Grand Duke of Tuscany, a place he occupied until his early death caused by typhoid fever.

Evangelista Torricelli
Portrait with a mercury barometer

Despite his remarkable work as a mathematician, Torricelli is especially known for being the first man to determine the atmospheric pressure, thanks to the barometer he designed based on a tube filled with mercury.

SOLUTION

First we will start by proving a similar theorem (about the sum of distances from a point to the sides of a triangle, and not to its vertices) and only for equilateral triangles. We will see later how we will rely on it to find the solution to our problem.

THEOREM 23.1. *(Viviani) In an equilateral triangle, the sum of the three distances from an interior point to the sides of the triangle is always the same, regardless of the position of the point.*

PROOF. Let us suppose an equilateral triangle with vertices P, Q and R, where g is the value of the length of a side, h is the value of the height of the triangle and A is the value of its area.

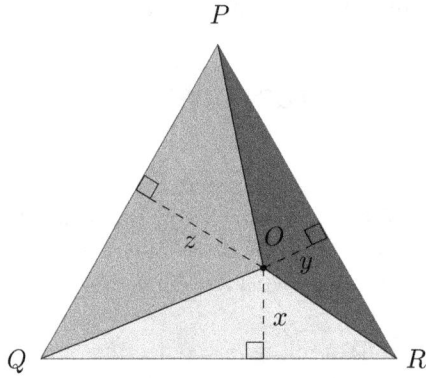

Figure 23.1

If the values x, y and z are the distances from an arbitrary point O to the sides $\overline{QR}$, $\overline{RP}$ and $\overline{PQ}$ (respectively), then the value of the area A of the triangle can be calculated as the sum of the areas of the three triangles drawn in figure 23.1:

$$A = \frac{1}{2}gx + \frac{1}{2}gy + \frac{1}{2}gz$$

Therefore, the sum of distances (S) is equivalent to the value $x + y + z$, and it can be calculated as:

$$S = x + y + z = \frac{2A}{g} = h$$

That is, S is independent of the position of point O and its value is equal to the height of the triangle. □

Taking advantage of this theorem, we will look for the solution to the original problem, although we will first do it only for triangles whose angles have all values less than 120°.

THEOREM 23.2. *(Fermat) For those triangles ABC whose angles have all values less than 120°, the point O whose sum of distances to the vertices is a minimum is the point where perpendiculars lines from A, B and C to segments $\overline{AO}$, $\overline{BO}$, $\overline{CO}$ (respectively) form an equilateral triangle.*

PROOF. Let us first see that such point O exists, that it is unique and that it is inside the triangle ABC. We build for each side of the triangle an arc where all subtended angles are equal

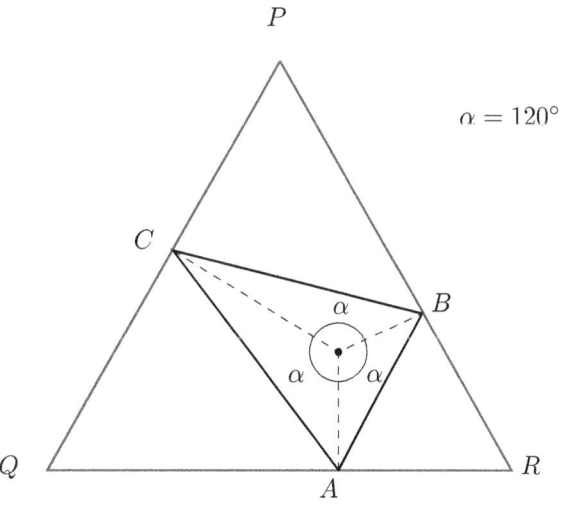

Figure 23.2

to 120° (see theorem 25.1). Let T be the point where the arcs build for $\overline{AB}$ and $\overline{BC}$ meet; then, the value of both angles $\widehat{ATB}$ and $\widehat{BTC}$ is equal to 120° (by the property of subtended angles), so the angle $\widehat{CTA}$ should also be equal to 120° in order to complete the value of 360°.

So the point T is also part of the arc built for $\overline{CA}$ and that means that the three arcs meet in a single point, that is T. The reasoning is valid only if all angles of triangle ABC have a value that is smaller than 120°, since otherwise the arc built for side $\overline{AB}$ (assuming that C is the vertex whose angle has a value greater than 120°) is outside the triangle and does not cut either of the other two arcs.

Therefore, point T is the same as point O described in the statement. Now, as we can see in figure 23.2, if $\widehat{AOB} = 120°$, $\widehat{OAR} = 90°$ and $\widehat{OBR} = 90°$, it follows that $\widehat{ARB} = 60°$. The same deduction for $\widehat{AQC}$ and $\widehat{CPB}$ leads to the equilateral triangle PQR we are looking for.

We want to prove that the sum of distances $\overline{AO} + \overline{BO} + \overline{CO}$ is minimum for any inner point of triangle ABC.

Let us take a look at figure 23.3, where O' is another point inside the triangle. If A', B' and C' are the perpendicular projections of point O' onto the sides of triangle PQR, then we have:

$$\overline{O'A'} \leq \overline{O'A} \qquad \overline{O'B'} \leq \overline{O'B} \qquad \overline{O'C'} \leq \overline{O'C}$$

since the values of the distances from points projected perpendicularly to the sides will always be smaller than (or equal to) the values of the distances from any other point to the same sides.

In fact, the equal sign in the first inequality (for example) is only achieved if A and A' are the same point, i.e., if A, O and O' are in the same line. If that happens with point A, it is not possible with B and C at the same time (if O and O' are different points); that is, the three equalities cannot be satisfied at the same time (in figure 23.3 we have chosen a point O', different from O, whose perpendicular to side $\overline{QR}$ is the same to the perpendicular from point O to the same side, so in this case only A and A' are the same point).

If we now sum the three previous inequalities (which CANNOT be satisfied all at the same time):

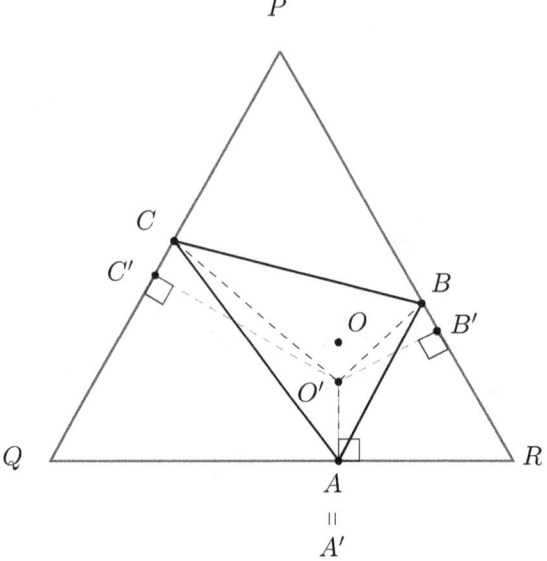

Figure 23.3

$$\overline{O'A'} + \overline{O'B'} + \overline{O'C'} < \overline{O'A} + \overline{O'B} + \overline{O'C}$$

However, by Viviani's theorem we have that the value $\overline{O'A'} + \overline{O'B'} + \overline{O'C'}$ is equal to the value of $\overline{OA} + \overline{OB} + \overline{OC}$ (the sum of distances to the sides of an equilateral triangle). Therefore, replacing the previous inequality:

$$\overline{OA} + \overline{OB} + \overline{OC} < \overline{O'A} + \overline{O'B} + \overline{O'C}$$

$\square$

It only remains to study what happens if an angle of the triangle has a value that is greater than 120° (obviously, there cannot be more than one angle with this property).

Observation: In case that a value of an angle is equal or greater than 120° (let us suppose that is the angle in vertex C), then the point solution is precisely the vertex C. Specifically, in this case we have that $\overline{AC} + \overline{BC} < \overline{AU} + \overline{BU} + \overline{CU}$, regardless of where the point U is.

PROOF. Let γ be the value of the angle $\widehat{ACB}$ and let F (resp., G) be the projection of the point U onto the line that includes the side $\overline{AC}$ (resp.,$\overline{CB}$).

We have to consider three different cases, depending on where the point U is inside the triangle ABC (see figure 23.4): 1) Both points F and G are in the segments $\overline{AC}$ and $\overline{CB}$, 2) Point G is in the segment $\overline{CB}$, but point F is not in the segment $\overline{AC}$, and 3) Point F is in the segment $\overline{AC}$, but point G is not in the segment $\overline{CB}$.

Let x (resp., y) be the value of the distance between C and F (resp., G), which we consider a positive value if F is in the segment $\overline{AC}$ (resp., $\overline{CB}$) and negative otherwise. Finally, let ψ (resp., ϕ) be the value of the angle $\widehat{UCF}$ (resp., $\widehat{UCG}$), considering that the value of the angle is positive if x is positive, and negative otherwise (i.e, if x is negative).

In any of these three cases:

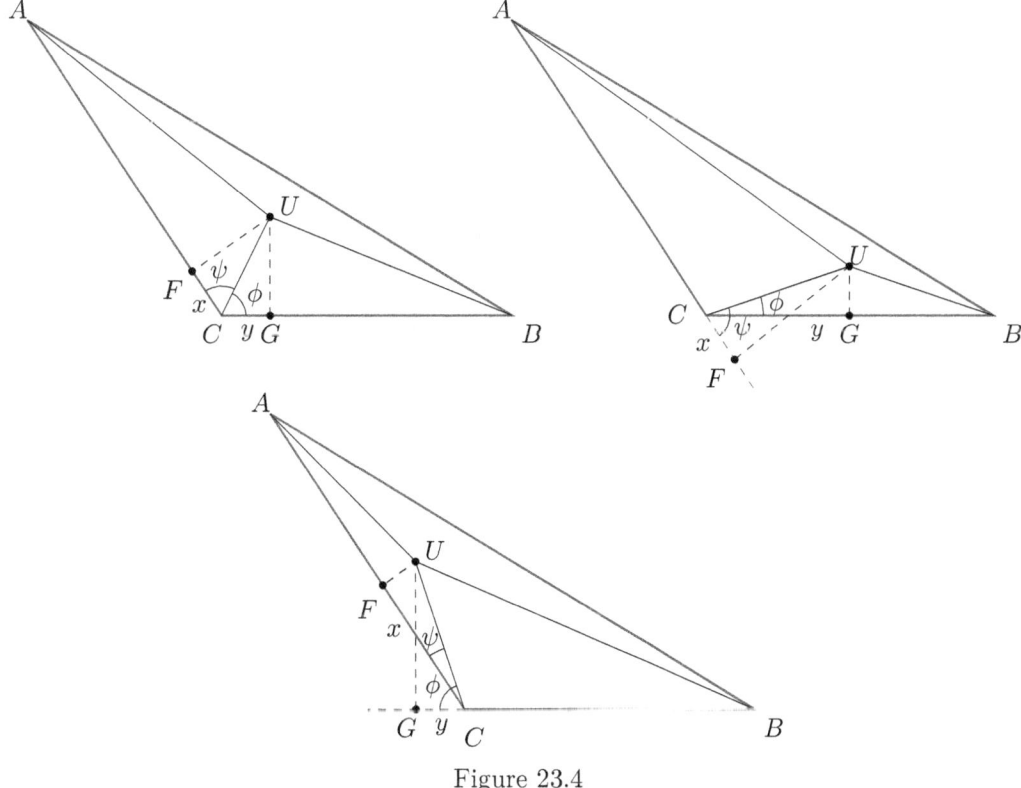

Figure 23.4

$$x = \overline{CU} \cdot \cos\psi \qquad y = \overline{CU} \cdot \cos\phi$$

considering the signs of all the variables, which also leads us to the following equation (valid in all three cases):

$$\overline{AC} = \overline{AF} + x \qquad \overline{BC} = \overline{BG} + y$$

Consequently:

(23.1) $$\overline{AC} + \overline{BC} = \overline{AF} + \overline{BG} + x + y$$

Now:

$$x + y = \overline{CU}\cos\psi + \overline{CU}\cos\phi = CU \cdot (\cos\psi + \cos\phi) =$$

(23.2) $$= 2 \cdot \overline{CU} \cdot \cos\left(\frac{\psi + \phi}{2}\right) \cdot \cos\left(\frac{\psi - \phi}{2}\right)$$

In the first case, we have that $\psi + \phi = \gamma$, from where it follows that $(\psi + \phi)/2 \geq 60°$ (by hypothesis, $\gamma \geq 120°$) and, therefore, $|\cos((\psi + \phi)/2)| \leq 1/2$, which implies, in equation (23.2), that $x + y \leq \overline{CU}$.

143

In the other two cases we have instead that $|\psi - \phi| = \gamma$, from which it is now deduced that $|(\psi - \phi)/2| \geq 60°$. But now it s true that $|\cos((\psi - \phi)/2)| \leq 1/2$, which implies again that $x + y \leq \overline{CU}$ (although now due to another term in equation (23.2)).

So it is true that $x + y \leq \overline{CU}$ in any of the three cases, which substituted in (23.1) leads us to:

$$\overline{AC} + \overline{BC} \leq \overline{AF} + \overline{BG} + \overline{CU}$$

Finally, as the sides $\overline{AF}$ and $\overline{BG}$ of the right triangles AUF and BUG are shorter than the hypotenuses AU and BU (except if point U is the same as point C), then:

$$\overline{AC} + \overline{BC} \leq \overline{AU} + \overline{BU} + \overline{CU}$$

which completes the proof. $\qquad\qquad\qquad\qquad\qquad\qquad\qquad\qquad\qquad\qquad\qquad$ □

We have reached the solution:

- If the original triangle has an angle whose value is greater than or equal to 120°, the sought point is the vertex corresponding to this angle.

- If all three angles of the original triangle have values that are smaller than 120°, the sought point is the intersection of the arcs built on the three sides of the triangle where subtended angles have a value equal to 120°. This intersection point always exists and it corresponds to an interior point of the triangle.

FINAL REMARKS

A possible practical application of this problem would be to use that point for the construction of a railway line or a fiber optic connection that connects three large cities in a region, with the goal of using as little material as possible, or to build a shopping center close to these three cities trying to attract the maximum number of clients (although in some real cases the promoters of the project mistakenly believed that the ideal point was the barycenter of the imaginary triangle that joined the three cities).

Chapter 24

The astroid

(Leibniz – 1652)

PROBLEM

To calculate the envelope of a fixed length segment whose end points slide through the coordinate axes.

HISTORY

Let us suppose that we have a segment of length L whose end points A and B are originally at points $(0, L)$ and $(0, 0)$ of the coordinate axes. Let us imagine this as a ladder that it is on the ground, supported by two walls (we are watching the scene from "above"). We can move the ladder slowly, with the condition that both its end points are always on the walls, until a final position where the end points of the ladder are at points $(0, 0)$ and $(L, 0)$, as we can see in the first graph of figure 24.1.

If we think about the position of all points of the ladder (in all intermediate positions), we are creating an area whose border is the desired envelope, as defined by the mathematicians of antiquity. We can see the resulting area and its envelope in the second graph of figure 24.1.

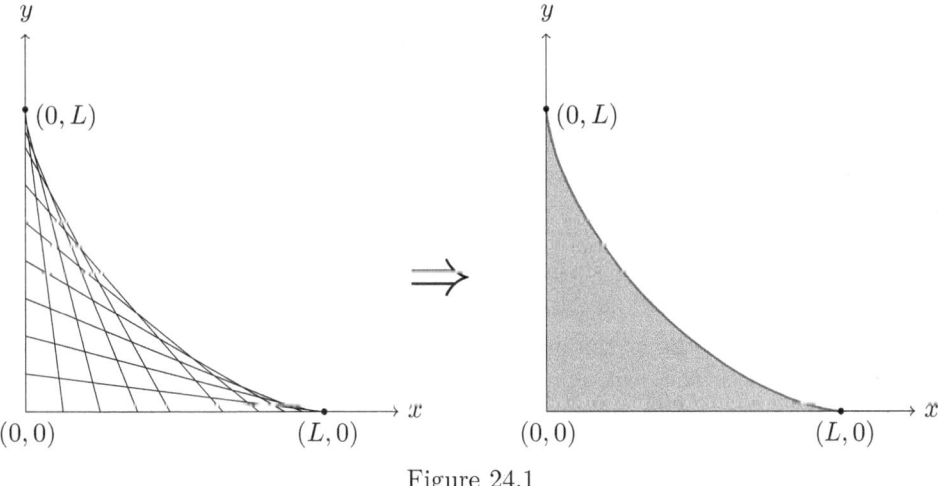

Figure 24.1

The German mathematician Gottfried Wilhelm LEIBNIZ (1646 – 1716) founded the theory of envelopes in 1692. Other mathematicians have found interesting applications of envelopes in optics or differential equations. The problem that concerns us will be solved with simple mathematics and a good dose of ingeniousness.

GOTTFRIED WILHELM LEIBNIZ 1646-1716 DEUTSCHLAND

German stamp commemorating Leibniz

SOLUTION

First approximation

Let us slightly change the initial approach to the problem and think that, in addition to the initial walls, there are other walls so that there are two corridors of width a and b with a corner at the origin of coordinates, as we can see in figure 24.2. In this case, what is the maximum length that the ladder can have to pass through the corner?

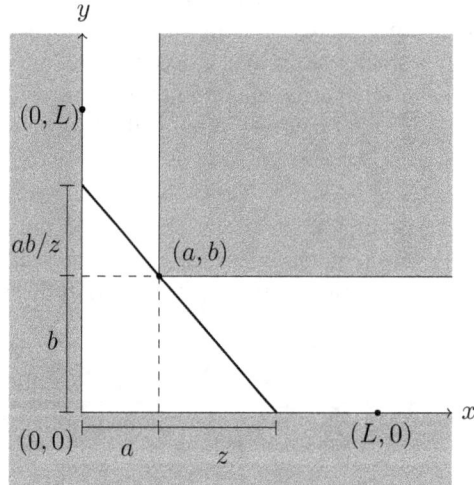

Figure 24.2

The problem can be treated as searching for an extreme of a one-variable function. Let (a, b) be the edge of the corner, where the "inner" walls meet, and let $(0,0)$ be where the "outer" walls (that are also the axes of the coordinate system) meet.

Let z be the distance between the end point of the ladder that lies on the axis X and the point $(a, 0)$. Then, applying Thales' theorem we have that the distance from the other end point of the ladder to the point $(0, b)$ is equal to (ab/z).

For every value of z we have a different length, we can think of a ladder of length $L(z)$ that touches the vertex of the corner (a, b) and whose end points are in the outer walls (as in figure 24.2). Among all possible lengths $L(z)$, the smallest one has a special property: any ladder with a length that is bigger than this value, let us call it $L(z_0)$, cannot pass through the corner. Therefore, for the value z_0 we have the maximum length of the ladder $L(z_0)$ that can turn the bend; if we take any other other value of z, the ladder of length $L(z)$ would be longer that $L(z_0)$ and it is not possible to turn the corner with it.

By the Pythagorean theorem we know that:

$$L^2(z) = \left(b + \frac{ab}{z}\right)^2 + (a+z)^2 = \cdots = \left(\frac{b^2}{z^2} + 1\right) \cdot (a+z)^2$$

We want to minimize this function (it is the same to minimize a value than to minimize its square), so we can calculate its derivative and equal it to 0.

$$[L^2(z)]' = \left(-\frac{2b^2}{z^3}\right) \cdot (a+z)^2 + \left(\frac{b^2}{z^2} + 1\right) \cdot 2(a+z) = 0 \quad \Rightarrow \quad \cdots \quad \Rightarrow \quad z^3 = ab^2$$

You can verify that this extreme is a minimum so, in our previous notation:

$$z_0^3 = ab^2 \quad \rightarrow \quad z_0 = a^{1/3} b^{2/3}$$

Substituting

$$z_0 = a^{2/3} b^{1/3} \qquad \text{in} \qquad L^2(z) = \left(\frac{b^2}{z^2} + 1\right) \cdot (a+z)^2$$

we have that:

$$L^2(z_0) = \left(\frac{b^2}{a^{2/3} b^{4/3}} + 1\right) \cdot (a + a^{1/3} b^{2/3})^2 = \left(\frac{b^{2/3}}{a^{2/3}} + 1\right) \cdot a^{2/3} \cdot (a^{2/3} + b^{2/3})^2 =$$

$$= \left(\frac{a^{2/3} + b^{2/3}}{a^{2/3}}\right) \cdot a^{2/3} \cdot (a^{2/3} + b^{2/3})^2 = (a^{2/3} + b^{2/3})^3 \quad \Rightarrow \quad (L(z_0))^{2/3} = a^{2/3} + b^{2/3}$$

That is, the length we are looking for, let us call it L, meets the curious symmetric equation:

$$L^{2/3} = a^{2/3} + b^{2/3}$$

Second approximation

What can we deduce from the previous problem? First, once we have calculated the longest length of the ladder that can go through the corner, we know that the time with greatest difficulty is precisely when the equation $z^3 - ab^2$ is satisfied. For the rest of the ladder positions, as shown in figure 24.3, we will have enough space (the other positions allowed a longer length of a ladder, since the minimum is reached for z_0).

To understand figure 24.3: let us consider a segment that starts at a point on the axis x other than $(a + z_0, 0)$, passes through the point (a, b) and ends at the axis y. This segment has a length whose value is greater than L, since the minimum length was obtained when we have one end point of the ladder at $(a + z_0, 0)$. So if the ladder of length L does not have an end point at the x axis other than $(a + z_0, 0)$, then it does NOT touch any point of the inner walls.

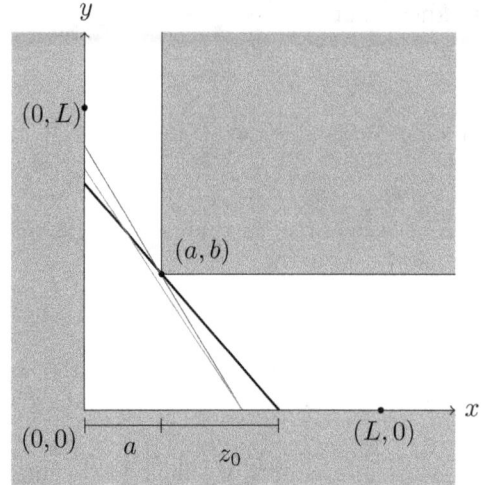

Figure 24.3

That is, the point (a, b) satisfies that it is a point **of the envelope** of a ladder of length L, since it is a point of the ladder in a certain position, but for the rest of positions of the ladder it is never reached (therefore, it is a point of the border of the area of Figure 24.1).

If we took another point (a', b') that satisfied

$$a^{2/3} + b^{2/3} = (a')^{2/3} + (b')^{2/3}$$

and we raised again the problem of the ladder and the corner, we would reach the same conclusion. In this case, the maximum length of the ladder would be:

$$L^{2/3} = a^{2/3} + b^{2/3} = (a')^{2/3} + (b')^{2/3}$$

and the point (a', b') is a point of the envelope.

Reasoning in the same way, all points with positive coordinates (x, y) that satisfy the equation

$$L^{2/3} = x^{2/3} + y^{2/3}$$

are points of the envelope, while other points are not. We have obtained the solution of the problem: the envelope is the graph of the function that satisfies the equation $L^{2/3} = x^{2/3} + y^{2/3}$.

If we draw the curve for all four quadrants of the system of coordinates (the problem only makes sense in the first quadrant, but the equation has a solution for negative values of x or y, too), the result is the graph in figure 24.4.

The mathematician J.J. LITTROW thought that graph was similar to a star ("astron" in Greek), so he proposed **astroid** as its name and it has lasted to present.

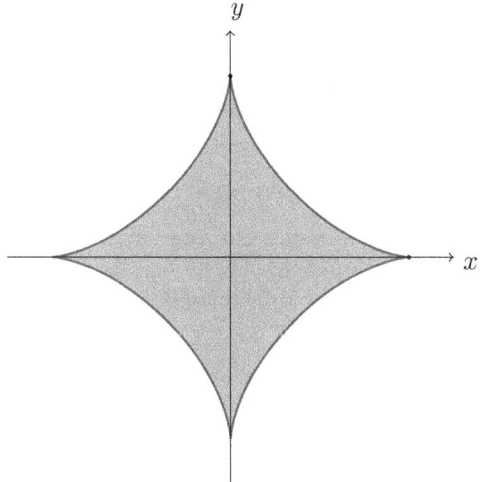

Figure 24.4

FINAL REMARKS

- Although in this problem we have sought a reasoned solution with more basic procedures, Leibniz found a generic method to find the envelope of a family of curves with equations $F_t(x, y) = 0$. In our example, these curves are the family:

$$\frac{x}{t} + \frac{y}{\sqrt{l^2 - t^2}} - 1 = 0 \qquad \text{for} \qquad 0 < t < l$$

According to Leibniz's theory, the envelope is the solution of the system of equations:

$$\begin{cases} F_t(x, y) = 0 \\ \dfrac{\delta F_t(x, y)}{\delta t} = 0 \end{cases}$$

that is, the intersection of the family of curves and the family of its partial derivatives with respect to the parameter t.

In our example:

$$\begin{cases} F_t(x, y) = 0 & \Rightarrow & \dfrac{x}{t} + \dfrac{y}{\sqrt{l^2 - t^2}} - 1 = 0 \\ \dfrac{\delta F_t(x, y)}{\delta t} = 0 & \Rightarrow & \dfrac{-x}{t^2} + \dfrac{-y}{2\sqrt{(l^2 - t^2)^3}} \cdot (-2t) = 0 \end{cases}$$

Isolating the value of t in the second equation

$$t = \frac{l \cdot x^{1/3}}{\sqrt{x^{2/3} + y^{2/3}}}$$

and replacing it in the first equation, we get the same equation of the astroid $l^{2/3} = x^{2/3} + y^{2/3}$ that we saw in the main solution of the problem.

149

- Despite appearing as a result of a problem with ladders and corners, the astroid fulfills a curious property that is not related with this origin: it is also the hypocycloid resulting from rolling a circle of radius r along the border inside a (larger) circle of radius equal to $4r$.

By definition, if a circle C_1 rolls along the border of another circle C_2, a fixed point of the first one describes an **epicycloid** when C_1 rolls along the border outside C_2, and it describes an **hypocycloid** when it rolls inside C_2. We will see the definition of cycloid in the problem "The perfect pendulum".

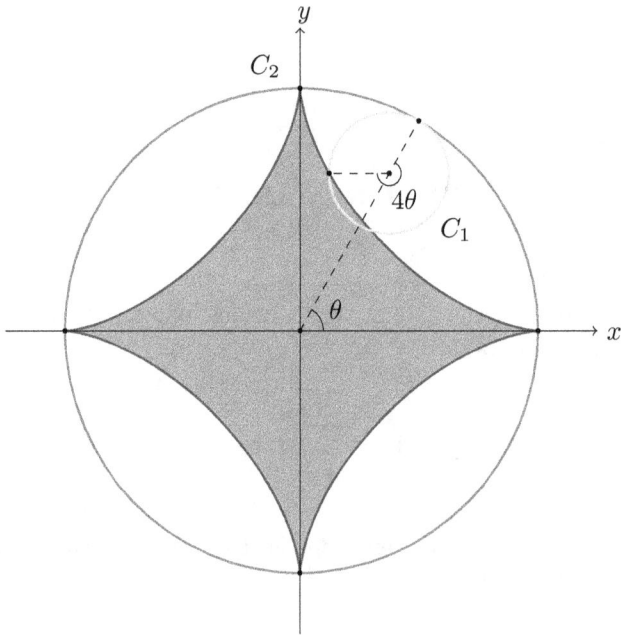

Figure 24.5

In figure 24.5, the point A of the circle C_1 (of radius r) describes an astroid when C_1 rolls inside of C_2 (of radius $4r$). The proof of this property is left to the reader.

Chapter 25

The sliding triangle

(Van Schooten – 1657)

PROBLEM

To find the path that the vertex of a triangle describes when the other two vertices slide through two non-parallel lines (each vertex point through a different line).

HISTORY

Frans van Schooten (1615 – 1660) was a Dutch mathematician who is mainly remembered for rewriting in a didactic way the books of Descartes, who he met personally. He was also one of the main promoters of his Cartesian geometry, publishing the first Latin version of the work "La géométrie" (1649).

Frans van Schooten

Impressed by Descartes' method, he wrote his own books based on his work, composed of plenty of interesting problems like the one we are dealing with. Some of these problems are about trajectories of points (or "locus", in mathematical language) and they can be useful for mechanics or engineering. Specifically, this problem appears in the book "Exercitationes mathematicae" (1657).

SOLUTION

Before trying to solve the problem we will look for partial results that will be useful to us. We will start with the geometrical theorem of the "subtended angles" (used in other problems of this book) and its interesting properties.

THEOREM 25.1. *(Subtended angles) Let $\overline{AB}$ be a chord of the circumference C (with center in O). Then it is true that (see figure 25.1):*

a) The angle $\widehat{APB}$ (where P is a point of the arc of circumference AB – specifically, the arc of shortest length between the two existing arcs that join A and B) has always the same value, *regardless of the chosen point P. Let α be that angle. In general $180° > \alpha \geq 90°$, the equality being held when $\overline{AB}$ is a diameter of C.*

b) Any other point P' (that lies in the same half plane of P of those separated by the line containing $\overline{AB}$) that does not belong to C meets the property that the value of the angle $\widehat{AP'B}$ is different to α (it is greater if P' is inside C, and it is smaller if it is outside).

c) The angle $\widehat{AQB}$ (where Q is a point of the arc AB – now the one of longest length between the two of them) has always the same value, regardless of the chosen point Q. Let β be that angle. Then, $\beta = 180° - \alpha$, which implies that $0 < \beta \leq 90°$ and that $\beta \leq \alpha$ (equalities being held only if $\overline{AB}$ is a diameter of C)

d) Any other point Q' (that lies in the same half plane of Q of those formed by the line containing $\overline{AB}$) that does not belong to C meets the property that the value of the angle $\widehat{AQ'B}$ is different to β (it is greater if Q' is inside C, and it is smaller if it is outside).

e) The value of the angle $\widehat{ABO}$ is equal to $\alpha - 90°$.

f) The value of the angle $\widehat{AOB}$ (central angle) is double than the value of the angle β.

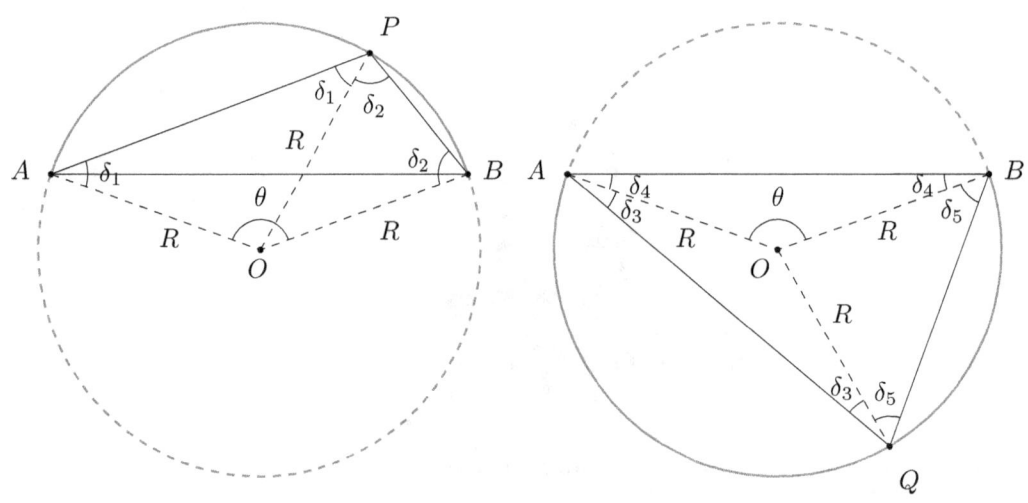

Figure 25.1

PROOF. In the left graph of figure 25.1, since P is a point of the circumference (it does not matter which one it is, as long as it lies on the arc AB of shorter length - that is, the one drawn with continuous stroke), it is inferred that the length of the segment $\overline{OP}$ corresponds to the radius R of the circumference.

Then, the triangles OAP and OPB are isosceles, which implies the equalities of the angles shown in the figure ($\widehat{OAP} = \widehat{OPA} = \delta_1$ and $\widehat{OBP} = \widehat{OPB} = \delta_2$). Furthermore, the four angles of the quadrilateral $OAPB$ must sum $360°$ (the sum of the angles of two triangles), which leads us to the equation $2 \cdot (\delta_1 + \delta_2) + \theta = 360°$, i.e., $\alpha = (\delta_1 + \delta_2) = 180° - \theta/2$, proving section (a). Also, in the last equation it follows that if $\theta = 180°$ (that is, the chord $\overline{AB}$ is a diameter of the circle) then $\alpha = (\delta_1 + \delta_2) = 90°$.

In the right graph of figure 25.1, since Q is a point of the circumference (now it lies on the arc AB of greater length - again drawn with continuous stroke), it is deduced that the length $\overline{OP}$ corresponds to the radius R of the circumference.

Then, the triangles OAQ, OAB and OQB are isosceles, which implies the equalities of the angles shown in the figure ($\widehat{OAQ} = \widehat{OQA} = \delta_3$, $\widehat{OAB} = \widehat{OBA} = \delta_4$ and $\widehat{OBQ} = \widehat{OQB} = \delta_5$). Furthermore, the three angles of the triangle AQB must sum 180°, which leads us to the equation $2 \cdot (\delta_3 + \delta_4 + \delta_5) = 180°$, and taking into account that $2\delta_4 + \theta = 180°$ (triangle AOB) it follows that $\beta = (\delta_3 + \delta_5) = \theta/2 = 180° - \alpha$, proving sections (c) and (f).

Finally, from the aforementioned equation $2\delta_4 + \theta = 180°$ it follows that $\delta_4 = 90° - \theta/2 = \alpha - 90°$ (section e).

We leave to the reader the (easy) proofs of sections (b) and (d). □

Once we have studied the theorem of subtended angles, we come back to our original problem, starting with a very particular case: when the triangle is degenerate, that is, its three vertices are on the same line and the lines on which two of them slide through are perpendicular to each other (see figure 25.2). This result was already known in the fifth century BC.

PROPOSITION 25.1. *Let A, B, C be three points on a line. Suppose that A slides through a straight line r and that B slides through another straight line s that is perpendicular to the first one, so that the lengths between A, B and C remain constant. Let $a = |BC|$, $b = |AC|$ and $c = |AB|$, where $c = |a \pm b|$ and the sign depends on whether C lies between A and B or not. Then, point C describes the trajectory of an ellipse.*

PROOF. According to the notation used in figure 25.2, let δ be the angle between lines ABC and r. Suppose that r and s are the coordinate axes and let (x, y) be the variables that define the position of C in this coordinate system. It is then satisfied that:

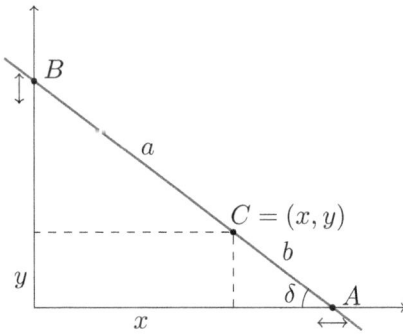

Figure 25.2

$$\begin{cases} x - a \cdot \cos \delta \\ y = b \cdot \sin \delta \end{cases} \rightarrow \begin{cases} (x/a) = \cos \delta \\ (y/b) = \sin \delta \end{cases}$$

If we square and add both equations:

$$\frac{x^2}{a^2} + \frac{y^2}{b^2} = 1$$

This equation is an ellipse with axes on the axes of coordinates and semiaxes a and b, which is true irrespective of C lying between points A and B or not. □

We are ready to look for the general solution to the problem.

THEOREM 25.2. *Let ABC be a triangle. Suppose that the vertex A slides through a line r and that the vertex B slides through another line s, and let θ be the fixed angle between them. Then, the vertex C describes the trajectory of an ellipse.*

PROOF. Let S be the intersection point between lines r and s. For each position of the triangle ABC (while its vertices A and B slide through these lines) we can draw the (unique) circumference D that passes through A, B and S. The center M of this circumference has a fixed position with respect to A and B (since it is in the perpendicular bisector of segment $\overline{AB}$ and, according to property (e) of subtended angles, the angle $\widehat{ABM}$ has always the same value: $\widehat{ABM} = \theta - 90°$), so we can consider that M "moves" with triangle ABC as a "solid" (that is, it preserves angles and distances with the three vertices).

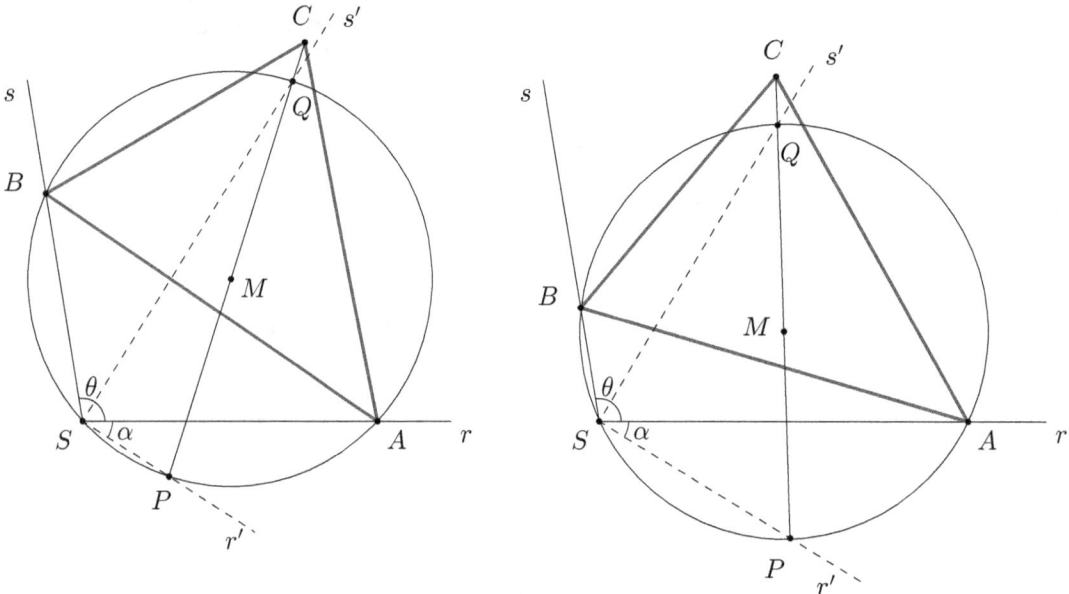

Figure 25.3

Now, for each position of the triangle (we see two of them in figure 25.3), let us join point M and vertex C with a line and let P and Q be the intersections of this line with circumference D. Since M moves as a solid with ABC, then points P and Q also move as a solid with the triangle, since the line $\overline{CM}$ moves as a solid with the triangle and the distances $|MP|$ and $|MQ|$ have always the same value (equal to $|MA|$ or $|MB|$, the radius of D, which is constant as we proved in the previous section).

Therefore, the segment $\overline{BP}$ (for example) has always the same length (for each position of the triangle ABC when its vertices A and B are sliding through lines r and s). For the property (e) of subtended angles, that means that the angle $\widehat{BSP}$ has always the same value $\widehat{BSP} = \widehat{PBM} + 90°$ (S is the only point that does not move as a solid with the triangle but lies always, by construction, in the circumference D), which forces the angle $\widehat{ASP}$ to be also constant (since $\widehat{ASB}$ is constant too, and $\widehat{BSP} = \widehat{PSA} + \widehat{ASB}$). That is, the point P moves when the triangle slides, but it always lies on a straight line r' that starts at point S.

The reasoning for point Q is identical, so we come to the same conclusion: the point Q moves when the triangle slides, but it always lies on a straight line s' that starts at points S.

Now, since the segment $\overline{PQ}$ is a diameter of the circumference D, the angle $\widehat{PSQ}$ is, for the property (a) of subtended angles (in the particular case of the equality), equal to $90°$.

In short, we have three points P, Q and C that lie on the same line, and two of them, P and Q, slide through two perpendicular lines. This situation is precisely the one we saw in proposition 25.1, so we can infer that the third point, C, follows the trajectory of an ellipse, with the semiaxes (of values the distances $|CP|$ and $|CQ|$) in lines r' and s'. $\qquad\square$

FINAL REMARKS

Before the appearance of computer programs that draw mathematical figures, this result was used to draw ellipses: if we managed to construct a ruler in such a way that two of its points slide through two lines that are perpendicular between them, and we made a small hole at an intermediate point of the ruler where we put a pencil then we draw an ellipse when moving the ruler (the distances of the semiaxes would be according to the distances from the hole to the end points of the ruler).

Chapter 26

The perfect pendulum

(Huygens – 1657)

PROBLEM

To build a pendulum whose period of oscillation does not depend on its amplitude of oscillation.

HISTORY

In the problem "Calculation of the position at the sea" (in the second part of this book) we will explain the difficulty that the ships faced for centuries in order to know one of the geographical coordinates (the longitude) at a certain moment of a journey (the latitude can be easily calculated). We will see there that the best way to solve the problem is to build a "chronometer" that can accurately calculate the time elapsed since the last time at which the longitude was known (for example, when the ship was docked in port).

Clocks based on a simple pendulum were not a solution to the problem, since any sudden movement of the ship (produced by waves during a storm, for example) caused a variation of the amplitude of oscillation of the pendulum and, therefore, of its period of oscillation. That made it impossible to calculate the elapsed time, since the mechanism counted the number of oscillations but each of them indicated a different amount of time.

Christian Huygens

The Dutch astronomer, physicist and mathematician Christian HUYGENS (1629 – 1695) found a theoretical solution to the problem while studying a curve called cycloid, as he wrote in his work "Horologium oscillatorium" (1673):

The simple pendulum cannot be considered to be a safe and uniform measure of time, because wide oscillations take longer than shorter oscillations; with the help of geometry I have found a method,

which was unknown before, of suspending the pendulum and I have investigated the curvature of
a certain curve that lends itself admirably to achieve the desired uniformity. Once I have applied
this form of suspension to the clocks, its operation became so smooth and safe that after numerous
experiences on land and on sea it is unquestionable that these clocks offer the greatest safety to
astronomy and navigation. The trajectory is the same that a nail stuck in a wheel that rolls on a
straight line describes in the air; mathematicians call it a cycloid and it has been carefully studied
because it has many other properties. But I have studied it for its application to the measure of
time, which I discovered while studying it with a purely scientific interest and without suspecting
the result.

Huygens built (theoretically) a pendulum whose period of oscillation was always the same, regardless of its amplitude of oscillation. This way, although waves varied the pendulum's trajectory,
each oscillation calculated the same amount of time and, therefore, an oscillation counter was
sufficient to accurately measure the longitude of the ship's position. In the solution we will see
how Huygens devised his "perfect pendulum".

SOLUTION

DEFINITION 26.1. *The cycloid is the curve generated by a point P belonging to a circle of radius
r (called "generatrix") when it rolls along a straight line (called "directrix") without sliding.*

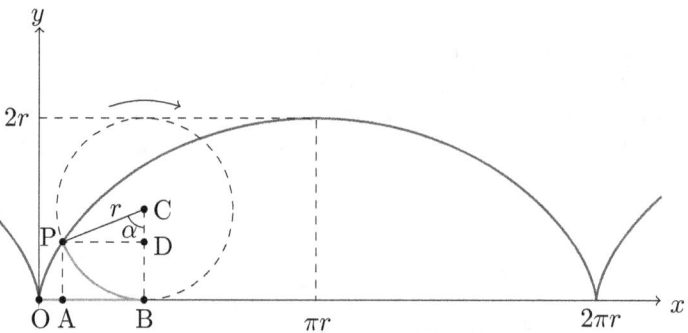

Figure 26.1

As we can see in figure 26.1, the cycloid is formed by several chained arcs. If we assume that
the circumference rolls on the X axis and that point P was initially at the origin of coordinates,
its trajectory passes through the point $(\pi r, 2r)$, where it reaches a maximum y value, and again
touches the X axis at point $(2\pi r, 0)$.

If B is the point of contact between the wheel and the X axis at any given moment, then the
horizontal distance $\overline{OB}$ is equal to the length of arch PB (this is the condition of rolling **without
sliding**). So when the point P reaches again the X axis it is at distance $2\pi r$ from the origin.

LEMMA 26.1. *The coordinates (x, y) of the points of the first arc of the cycloid can be written as
a function of a parameter α (equal to the angle $\widehat{PCB}$, where C is the center of the circumference)
in the following way:*

(26.1)
$$\begin{cases} x = r \cdot (\alpha - \sin \alpha) \\ y = r \cdot (1 - \cos \alpha) \end{cases} \qquad \alpha \in [0, 2\pi]$$

158

PROOF. In figure 26.1, the point A is the projection of point P onto the X axis and point D is the projection of point P onto the segment $\overline{BC}$. The x coordinate of point P is equal to the distance $|OB|$ minus the distance $|AB|$; the first one is equal to the value of $r\alpha$ (the length of arc PB), while the second one is equal to the value of $r\sin\alpha$ (looking at the triangle PCD).

Meanwhile, the y coordinate of point P is equal to the distance $|BC|$ minus the distance $|CD|$; the first one is equal to the value of r and the second one is equal to the value of $r\cos\alpha$ (looking again at the triangle PCD). $\square$

The cycloid fulfills two very interesting properties that we are going to study as two separate propositions. The first one is very curious and it is probably the reason why this curve has become widely well-known in Mathematics.

PROPOSITION 26.1. *The cycloid is a curve that has the **tautochrone** property (from Greek words "tauto" = same and "chronos" = time): if we use an inverted cycloid arc as an slope (figure 26.2) and we consider that gravity acts on the direction of the Y axis, then a ball left at rest at a point of the cycloid will reach the lowest point of it after the same amount of time, regardless of the initial position of the ball (we do not consider air or surface friction).*

PROOF. Consider figure 26.2, which represents a cycloid-shaped slope (we inverted it to match with the idea of gravity acting "down"). Let A be the starting point of the movement, where we will leave the ball initially at rest, and let B be the end point of the movement at the vertex of the cycloid. In parametric form, the coordinates of A and B are:

$$\begin{cases} (x_0, y_0) = (r \cdot (\alpha_0 - \sin\alpha_0), r \cdot (1 - \cos\alpha_0)) \\ (x_1, y_1) = (\pi r, 2r) \end{cases}$$

that is, the parameter in A is $\alpha = \alpha_0$, and in B is $\alpha = \pi$. What we want to show is that the amount of time that the ball takes to go from A to B is always the same, i.e., it does not depend on the value of α_0.

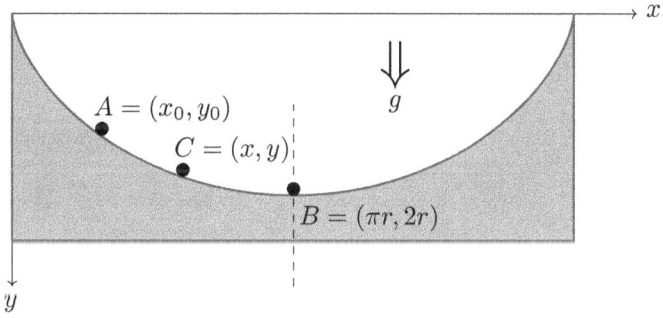

Figure 26.2

In order to see this property, let C be an intermediate point of the trajectory of the ball, with coordinates:

$$(x, y) = (r \cdot (\alpha - \sin\alpha), r \cdot (1 - \cos\alpha))$$

For the property of the conservation of energy (basically, the gravitational energy is transformed into kinetic energy; see details in the problem "The fastest slope" in the second part of this book), it is well known that the velocity of the ball at point C is:

$$v(\alpha) = \sqrt{2g \cdot (y - y_0)} = \sqrt{2gr \cdot (\cos \alpha_0 - \cos \alpha)}$$

Furthermore, the velocity of the ball at point C can be considered as a constant in a small interval of length close to it. Under this condition, this velocity of the ball quotient between the traveled distance and the amount of time it takes to travel it:

$$v(\alpha) = \frac{\Delta s}{\Delta t}$$

Both equations can be joined:

$$\Delta t = \frac{\Delta s}{\sqrt{2gr \cdot (\cos \alpha_0 - \cos \alpha)}}$$

This is the amount of time that the ball takes to travel a short distance close to point C; if we want to calculate the total amount of time from A to B then we must integrate this equation:

$$T(A \to B) = \int_A^B dt = \int_A^B \frac{ds}{\sqrt{2gr \cdot (\cos \alpha_0 - \cos \alpha)}}$$

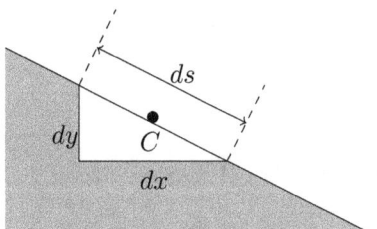

Figure 26.3

The usual method to treat the term ds in integral calculus is well known: from figure 26.3 we deduce that $ds = \sqrt{(dx)^2 + (dy)^2}$, but since the variables x, y are parametrized by α, we should better use this equation:

(26.2)
$$\frac{ds}{d\alpha} = \sqrt{\left(\frac{dx}{d\alpha}\right)^2 + \left(\frac{dy}{d\alpha}\right)^2}$$

As $dx/d\alpha$ and $dy/d\alpha$ are the derivatives of x, y with respect to parameter α, i.e.:

$$\frac{dx}{d\alpha} = r \cdot (1 - \cos \alpha) \qquad \frac{dy}{d\alpha} = r \sin \alpha$$

we can rewrite the equation 26.3 as:

$$\frac{ds}{d\alpha} = r \cdot \sqrt{2 - 2\cos\alpha} = r \cdot \sqrt{4 \cdot \frac{1 - \cos\alpha}{2}} = 2r \cdot \sqrt{(\sin\alpha/2)^2} = 2r \cdot \sin(\alpha/2)$$

Substituting in (26.2) we finally obtain an integral that only depends on the variable α:

$$T(A \to B) = \int_{\alpha_0}^{\pi} \frac{2r \cdot \sin(\alpha/2) \cdot d\alpha}{\sqrt{2gr \cdot (\cos\alpha_0 - \cos\alpha)}}$$

To solve it we apply again the trigonometric formula for the double angle:

$$\begin{cases} \cos\alpha_0 = 2\cos^2(\alpha_0/2) - 1 \\ \cos\alpha = 2\cos^2(\alpha/2) - 1 \end{cases}$$

to obtain:

$$\cos\alpha_0 - \cos\alpha = 2\cos^2(\alpha_0/2) - 2\cos^2(\alpha/2)$$

and therefore:

$$T(A \to B) = \sqrt{\frac{r}{g}} \cdot \int_{\alpha_0}^{\pi} \frac{\sin(\alpha/2) \cdot d\alpha}{\sqrt{\cos^2(\alpha_0/2) - \cos^2(\alpha/2)}}$$

We now apply the change of variable $\{\cos(\alpha/2) = t;\ -\sin(\alpha/2)/2 \cdot d\alpha = dt\}$:

$$T(A \to B) = -2 \cdot \sqrt{\frac{r}{g}} \cdot \int_{\cos(\alpha_0/2)}^{0} \frac{dt}{\sqrt{\cos^2(\alpha_0/2) - t^2}} = 2 \cdot \sqrt{\frac{r}{g}} \cdot \int_{0}^{\cos(\alpha_0/2)} \frac{dt}{\sqrt{\cos^2(\alpha_0/2) - t^2}}$$

and now we do the same with the change of variable $\{\cos(\alpha_0/2) \cdot u = t;\ \cos(\alpha_0/2) \cdot du = dt\}$:

$$T(A \to B) = 2 \cdot \sqrt{\frac{r}{g}} \cdot \int_{0}^{1} \frac{du}{\sqrt{1 - u^2}} = 2 \cdot \sqrt{\frac{r}{g}} \cdot [\arcsin u]_0^1 = \pi \cdot \sqrt{r/g}$$

Surprisingly, the value α_0 has disappeared from our calculations; the amount of time that the ball takes to reach point B is always equal to the value $\pi \cdot \sqrt{r/g}$, regardless of which point A we initially use as the starting position. $\qquad\square$

The second property of the cycloid that is necessary to study for the resolution of the problem was discovered by chance by Huygens, as we have said in the introduction to the problem.

PROPOSITION 26.2. *(Huygens) Suppose a roof consisting of two consecutive inverted cycloid arcs (obtained from a generating circle of radius r), with the midpoint at the origin of coordinates; and a pendulum consisting of a rope of length 4r and a small ball at its end, which is hanging from the origin of coordinates as in figure 26.4 (gravity acts again in the positive direction of the Y axis). Suppose that we swing the pendulum so that some part of the rope touches the roof wall, as in figure 26.4. Then the ball at the end of the pendulum follows the path of another cycloid arc.*

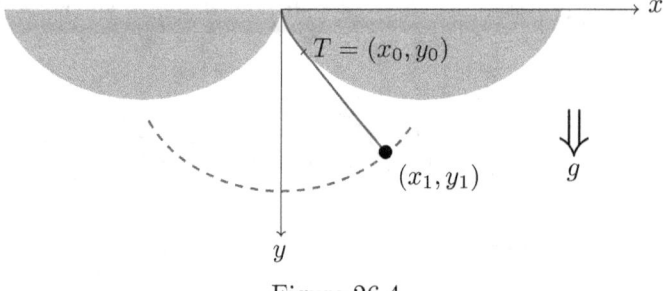

Figure 26.4

PROOF. Let us suppose that, in a certain moment of the oscillation, the coordinates of point T (last point of the rope in touch with the roof) are:

$$(x_0, y_0) = (r \cdot (\alpha_0 - \sin \alpha_0), r \cdot (1 - \cos \alpha_0)) \qquad \alpha_0 \in [0, \pi]$$

Using the standard notation, where ds is the differential element of length, we have that the length of the rope from $(0,0)$ to (x_0, y_0), following the cycloid border, is equal to:

$$L = \int_{(0,0)}^{(x_0, y_0)} ds$$

With the calculations of proposition 26.1, where we saw that for the cycloid of equation (26.1) the equality $ds = 2r \sin(\alpha/2) \cdot d\alpha$ is fulfilled, the resolution of the integral is easy:

$$L = \int_{(0,0)}^{(x_0, y_0)} ds = \int_0^{\alpha_0} 2r \sin(\alpha/2) \cdot d\alpha = 4r \cdot [-cos(\alpha/2)]_0^{\alpha_0} = 4r - 4r \cdot \cos(\alpha_0/2)$$

If the total length of the rope is $4r$ and the part that touches the roof has length equal to the value $4r - 4r \cdot \cos(\alpha/2)$, that means that the part of the rope that does **NOT** touch the roof has length equal to the value $4r \cdot \cos(\alpha/2)$. To see where the ball is at that moment (coordinates (x_1, y_1)) we must calculate therefore the point at a distance $4r \cdot \cos(\alpha/2)$ from point (x_0, y_0) in the direction of the **tangent** to the cycloid at that point.

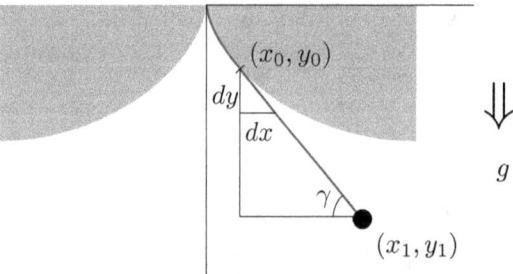

Figure 26.5

Since the tangent of a curve at a certain point coincides with its derivative dy/dx at that point (see figure 26.5), we have that coordinates (x_1, y_1) can be calculated as:

162

$$(26.3) \qquad \begin{cases} x_1 = x_0 + 4r \cdot \cos(\alpha_0/2) \cdot \cos\gamma \\ y_1 = y_0 + 4r \cdot \cos(\alpha_0/2) \cdot \sin\gamma \end{cases}$$

where

$$\tan\gamma = \frac{dy}{dx} = \frac{dy/d\alpha}{dx/d\alpha} = \frac{\sin\alpha_0}{1 - \cos\alpha_0}$$

Using trigonometric equations we infer that:

$$\cos\gamma = \sqrt{\frac{1 - \cos\alpha_0}{2}} \qquad \sin\gamma = \sqrt{\frac{1 + \cos\alpha_0}{2}}$$

and using the trigonometric equality of the half angle:

$$\cos(\alpha_0/2) = \sqrt{\frac{1 + \cos\alpha_0}{2}}$$

we can rewrite equation (26.3) as:

$$\begin{cases} x_1 = x_0 + 4r \cdot (\sin\alpha_0)/2 \\ y_1 = y_0 + 4r \cdot (1 + \cos\alpha_0)/2 \end{cases}$$

Finally, we substitute the values of:

$$(x_0, y_0) = (r \cdot (\alpha_0 - \sin\alpha_0), r \cdot (1 - \cos\alpha_0)) \qquad \alpha_0 \in [0, \pi]$$

and simplify the equation to obtain:

$$\begin{cases} x_1 = r \cdot (\alpha_0 + \sin\alpha_0) \\ y_1 = r \cdot (3 + \cos\alpha_0) \end{cases}$$

This calculation has been made assuming that point (x_0, y_0) is the last one that touches the roof, but the method would be identical if we take any other point. Therefore, the ball will follow a trajectory with coordinates:

$$(x, y) = (r \cdot (\alpha + \sin\alpha), r \cdot (3 + \cos\alpha)) \qquad \alpha \in [0, \pi]$$

What kind of curve is this, which starts at point $(0, 4r)$ for $\alpha = 0$ and ends at point $(\pi r, 2r)$ for $\alpha = \pi$? We will see the answer clearly if we make the change of variable $\alpha = \alpha' - \pi$; in this case, the equations are:

$$(26.4) \qquad \begin{cases} x = r \cdot (\alpha' - \pi + \sin(\alpha' - \pi)) = r \cdot (\alpha' - \sin \alpha') - \pi r \\ y = r \cdot (3 + \cos(\alpha' - \pi)) = r \cdot (1 - cos\alpha') + 2r \end{cases} \qquad \alpha \in [\pi, 2\pi]$$

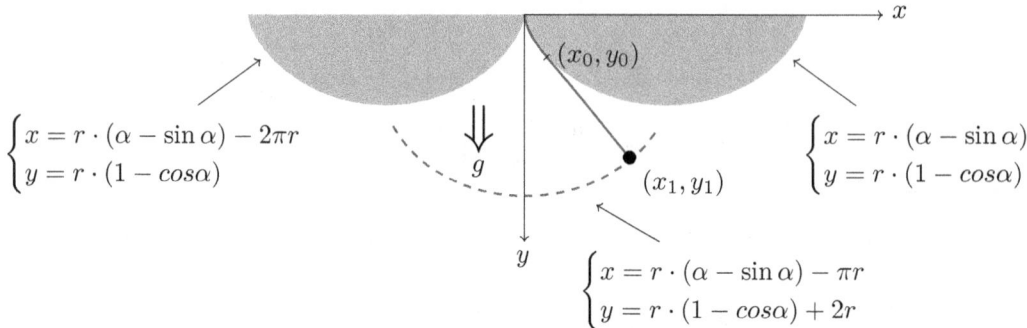

Figure 26.6

We can see that this is the equation of the second half of a cycloid arc, but displaced $-\pi r$ in the direction of the X axis and $2r$ in the direction of the Y axis, which coincides with the vertex at point $(0, 4r)$. By symmetry, when the pendulum is in the half plane $x < 0$, the ball will move according to the first half of the cycloid arc found. In short, the ball at the end of the rope is moving following the equation of a cycloid:

$$(x, y) = (r \cdot (\alpha' - \sin \alpha') - r\pi, r \cdot (1 - \cos \alpha') - 2r) \qquad \alpha' \in [0, 2\pi]$$

$\square$

The two propositions allow us to construct the "perfect pendulum": the pendulum described in proposition 26.2 follows a trajectory of a cycloid, which implies, by proposition 26.1, that it will always have the same period of oscillation, no matter from what point we initially release the ball and regardless of whether the amplitude of the oscillation change suddenly changes by some external movement. We have a pendulum that will always have an oscillation period equal to $4\pi \cdot \sqrt{r/g}$ (four times the calculated time in proposition 26.1, since there we calculate time from A to B , but the oscillation period starts at point A and it ends when it returns there again): it is the "perfect pendulum" of Huygens.

FINAL REMARKS

- In proposition 26.1 we consider a ball that rolls "with no friction" (and without inertia) along a slope with the shape of an inverted cycloid; in Proposition 26.2 we consider a ball at the end of a pendulum that follows a path with the shape of an inverted cycloid. Precisely because in both cases we do not suppose friction or inertia, we can deduce that both movements are equivalent and, therefore, affirm that the period of oscillation of the second case is constant, regardless of the amplitude of oscillation.

- Unfortunately for Huygens (and for the most powerful navies of that era), the "real" construction of the pendulum caused enormous problems. The friction of the rope with the cycloid-shape roof was not negligible and that was the reason why the clock did not work correctly. Huygens failed to put his marvelous theoretical discovery into practice, but he left this beautiful mathematical problem for posterity.

Chapter 27

Sum of squares

(Fermat – 1660)

PROBLEM

To prove that every prime number of the form $4n + 1$ can be represented, in a unique way, as the sum of two squares of natural numbers, and that no prime number of the form $4n + 3$ can be represented in that way.

HISTORY

Many statements in Number Theory, like this one that we are going to study, probably arose from the observation of curious mathematicians searching for coincidences. The French lawyer Pierre de Fermat (a mathematician in his spare time) was fascinated by looking for these properties, although he could not prove many of them (or, as in the case of the famous theorem that bears his name, he did not write the proof because the margin of the book he was reading was too small to do so).

In 1660 precisely Fermat noticed that all prime numbers of the form $4n + 1$ could be written as the sum of two squares ($5 = 1 + 4$, $13 = 4 + 9$, $17 = 1 + 16$, $29 = 4 + 25$, etc.) and that for each one there was only one way to do it. He also noticed that the same could not be done with any prime number of the form $4n + 3$ ($7 =?$, $19 =?$, $23 =?$, etc.).

Leonhard Euler
(old 10 swiss francs note)

He announced this result to the scientific community although he did not provide any proof. Nearly a century elapsed until Leonhard Euler wrote a first proof and he presented it in a treatise within the work "Novi Commentarii Academiae Petropolitanae ad annos 1754–1755", volume 5. Euler spent several years working on finding the solution.

Some readers may consider that this kind of problems, without an apparent practical application, do not deserve such a big effort, but the reality is that mathematical tools used to prove the result made the science evolve enormously.

SOLUTION

It is necessary to have knowledge of modular arithmetic to understand the proof shown below. It is assumed that the reader already has it.

Quadratic and non-quadratic residues

DEFINITION 27.1. *Let a be an integer and p an odd prime number, such that $a \neq 0$ (mod p). We say that a is a **quadratic residue** module p if there exists an integer x such that $x^2 \equiv a$ (mod p). Otherwise we will say that a is a **non-quadratic residue** module p.*

For example, 10 is a quadratic residue module 13 since $6^2 \equiv 10$ (mod 13), while 8 is non-quadratic residue module 13 since there is no value x such that $x^2 \equiv 8$ (mod 13).

Throughout the problem, let r be the integer number $r = (p-1)/2$ and when we refer to **residues** or **non-residues** it will be understood as quadratic and non-quadratic residues module p.

LEMMA 27.1. *Within the set $\{1, 2, 3, \cdots, p-1\}$ there are exactly r values that are residues and exactly r values that are non-residues. The r residues can be calculated as $1^2, 2^2, \cdots, r^2$ (mod p).*

PROOF. Let us first look at the next table, where the squares of $\{1, 2, 3, \cdots, p-1\}$ for $p = 13$ are, as that will give us an idea for the proof:

x	1	2	3	4	5	6	7	8	9	10	11	12
x^2 (mod 13)	1	4	9	3	12	10	10	12	3	9	4	1

There is a half-table symmetry that suggests the identity $x^2 \equiv (p-x)^2$ (mod p). But this is easy to prove, because $(p-x)^2 \equiv p^2 - 2px + x^2 \equiv x^2$ (mod p), since both p^2 and $-2px$ are multiples of p and, therefore, they are equal to 0 (mod p).

If we now show that the squares of the first half of the table ($\{1, 2, 3, \cdots, r\}$) are all different, the proof will be finished, since we will have r values that are residues (twice each), leaving r values that do not appear in the table as non-residues.

Let us suppose that $1 \leq x < y \leq r$ and that $x^2 \equiv y^2$ (mod p). In this case it follows that $y^2 - x^2 \equiv 0$ (mod p) $\Rightarrow (y+x) \cdot (y-x) \equiv 0$ (mod p). But $(y+x)$ can only take values between 3 (minimum value, when $x = 1$, $y = 2$) and $p-2$ (maximum value, when $x = r-1$, $y = r$); while $(y-x)$ can only take values between 1 (minimum value, when x and y are consecutive values) and $r-1$ (maximum value, when $x = 1$, $y = r$). That is, the product $(y+x) \cdot (y-x)$ is between two values that are different from 0 and that are not multiples of p, so it is impossible that the product is $\equiv 0$ (mod p), since p is a prime number. We have reached a contradiction so the assumption that $x^2 \equiv y^2$ (mod p) is not correct. $\square$

LEMMA 27.2. *The product of two residues or two non-residues is a residue, but the product of a residue and a non-residue is a non-residue.*

In the table provided for $p = 13$, we can see that both 4 and 12 are residues, and their product $4 \cdot 12 \equiv 48 \equiv 9$ (mod 13) is a residue too; similarly, both 5 and 8 are non-residues, and their product $5 \cdot 8 \equiv 40 \equiv 1$ (mod 13) is a residue; but the result of multiplying a residue and a non-residue, such as $4 \cdot 5 \equiv 20 \equiv 7$ (mod 13), is a non-residue. Let us prove all these results for any p.

PROOF. First, let x_1 and x_2 be two residues. Then, there exists a value y_1 such that $y_1^2 \equiv x_1$ (mod p) and there exists a value y_2 such that $y_2^2 \equiv x_2$ (mod p). Multiplying both equations we infer that $y_1^2 \cdot y_2^2 \equiv x_1 x_2$ (mod p) $\Rightarrow$ $(y_1 y_2)^2 \equiv x_1 x_2$ (mod p), so $x_1 x_2$ is a residue.

Now, suppose that the product of x_1 (residue such that $y_1^2 \equiv x_1$ (mod p)) and z_1 (non-residue) is a residue. In that case, we could write:

$$x_1 z_1 \equiv a^2 \ (\text{mod } p) \quad \Rightarrow \quad y_1^2 z_1 \equiv a^2 \ (\text{mod } p) \quad \Rightarrow$$
$$\Rightarrow \quad z_1 \equiv (y^{-1})^2 a^2 \ (\text{mod } p) \quad \Rightarrow \quad z_1 \equiv (y^{-1} \cdot a)^2 \ (\text{mod } p)$$

and we conclude that z_1 is a residue, which is impossible by hypothesis. Therefore, we have reached a contradiction, so the assumption that the product of x_1 and z_1 is a residue was wrong. It follows then that the product of a residue and a non-residue results in a non-residue.

Finally, let z_1 be a non-residue. If we multiply it by all the values of the set $\{1, 2, 3, \cdots, p-1\}$ we will obtain $p-1$ different values module p (well-known property of modular arithmetic), i.e., the set $\{1, 2, 3, \cdots, p-1\}$ again although in a different order. This set, by lemma 27.1, has r values that are non-residual (in the example of $p = 13$ these are the values $\{2, 5, 6, 7, 8, 11\}$) and, by the proof of the previous case, these values are the results of multiplying z_1 by each of the r residues; this means that the other r values resulting from the multiplications (i.e. residues) must be obtained when we multiply z_1 by each of the r non-residues. In summary, it follows that the multiplication of z_1 and a non-residue is a residue. $\square$

Is -1 a quadratic residue?

Obviously, the value of 1 is always a quadratic residue, regardless of the prime p that we take as a module. However, the value of -1 has a more unstable behavior: for some numbers p it is a residue, while for others it is a non-residue. For example, the value of $-1 \equiv 12 \equiv 5^2$ (mod 13) is a residue for $p = 13$, but it is not for $p = 11$ since there is no x such that $x^2 = -1 = 10$ (mod 11), as we can check in the following table:

x	1	2	3	4	5	6	7	8	9	10
x^2 (mod 11)	1	4	9	5	3	3	5	9	4	1

In this property a vital difference between prime numbers of the form $4n + 1$ and those of the form $4n + 3$ appears: in the first case, -1 is a residue, while in the second case it is a non-residue, as we will prove soon.

DEFINITION 27.2. *Let D be an integer number with no common divisor with p. We say that x and y are D-conjugates if $xy \equiv D$ (mod p).*

PROPOSITION 27.1. *Let D be an integer other than 0 module p. D is a residue if and only if it $D^r \equiv 1$ (mod p), while it is a non-residue if and only if $D^r \equiv -1$ (mod p).*

PROOF.

Case 1: D is a non-residue.

In this case we can group the elements of set $\{1, 2, 3, ..., p-1\}$ in pairs of D-conjugates, so that no element is repeated. For example, in the case $p = 13$ and for $D = 5$ (a non-residue), we have:

$$1 \cdot 5 \equiv 2 \cdot 9 \equiv 3 \cdot 6 \equiv 4 \cdot 11 \equiv 7 \cdot 10 \equiv 8 \cdot 12 \ (\text{mod } 13)$$

As D is a non-residue, there cannot be a number x that is conjugated of itself (since we would have $x^2 \equiv D \pmod{p}$ and D would be a residue) and, furthermore, the same number cannot appear in two groups of different conjugates since $ab_1 \equiv ab_2 \pmod{p}$ cannot be true (for p prime) if a is different from 0 module p and b_1 and b_2 are different values module p.

If we now multiply all the elements of the set $\{1, 2, 3, ..., p-1\}$ we see that, grouping them in pairs of conjugates, we have $(p-1)/2 = r$ times the value of D, that is:

$$(27.1) \hspace{3cm} (p-1)! \equiv D^r \pmod{p}$$

Case 2: D is a residue.

In this case we have to be a little more careful, because we have two values such that their square is equal to D module p (as we saw in lemma 27.1, because D is a residue): if we set x to one of them, the other one is then equal to $(p-x)$ and it is true that $x^2 \equiv (p-x)^2 \equiv D \pmod{p}$. The rest of the elements of the set $\{1, 2, 3, ...p-1\}$ can be grouped in pairs of conjugates, as in the previous case, with no more repetitions. For example, in the case $p = 13$ and for $D = 4$ (a residue), we have:

$$1 \cdot 4 \equiv 2 \cdot 2 \equiv 3 \cdot 10 \equiv 5 \cdot 6 \equiv 7 \cdot 8 \equiv 11 \cdot 11 \equiv 9 \cdot 12 \pmod{13}$$

Notice now that we have one more group in the equalities, since we have two special cases mentioned before (instead of one if they were grouped as a pair). Furthermore, when we multiply the values of x and $(p-x)$ we obtain $x \cdot (p-x) \equiv xp - x^2 \equiv -x^2 \equiv -D \pmod{p}$. Therefore, if we multiply all the values of the set $\{1, 2, 3, ..., p-1\}$ except x and (px) we can group them in conjugates to find $((p-1)/2) - 1 = r - 1$ times the value of D, while if we continue multiplying by x and $(p-x)$ we have to do it one more time by $-D$, that is:

$$(27.2) \hspace{3cm} (p-1)! \equiv D^{r-1} \cdot (-D) \equiv -D^r \pmod{p}$$

Summary: If we substitute D by 1 (value that trivially is always a residue regardless of the value of p) in expression (27.2) we find out that $(p-1)! \equiv -1 \pmod{p}$, and this result is known as Wilson's theorem. Now, if we substitute $(p-1)!$ by -1 in equation (27.1) and in equation (27.2), we finally obtain what we want to prove:

$$\begin{cases} -1 \equiv D^r \pmod{p} & \Rightarrow \quad D^r \equiv -1 \pmod{p} \quad \text{if } D \text{ is a non-residue} \\ -1 \equiv -D^r \pmod{p} & \Rightarrow \quad D^r \equiv 1 \pmod{p} \quad \text{if } D \text{ is a residue} \end{cases}$$

$\square$

For example, if we want to know whether $D = 5$ is a residue or not for $p = 13$ without building the complete table as in lemma 27.1, we should calculate the value of $5^6 \pmod{13}$. We would obtain that the result is equal to -1 and we would infer that 5 is a non-residue. If p is a large number, this method is viable (if we efficiently calculate the exponent operation), not so the calculation of the complete table.

COROLLARY 27.1. *The value -1 is a residue if and only if p is a prime number of the form $4n+1$, while it is a non-residue if and only if p is a prime number of the form $4n+3$.*

PROOF. If p is a prime number of the form $4n + 1$, we have that $r = (p - 1)/2 = 2n$ and therefore it is an even number. So when calculating $(-1)^r \equiv (-1)^{2n} \equiv 1 \pmod{p}$ we infer by proposition 27.1 that -1 is a residue.

However, if p is a prime number of the form $4n + 3$, we have that $r = (p - 1)/2 = 2n + 1$ and therefore it is an odd number. So when calculating $(-1)^r \equiv (-1)^{2n+1} \equiv -1 \pmod{p}$ we infer by proposition 27.1 that -1 is a non-residue. $\qquad\square$

Proof of the theorem for prime numbers of the form $4n + 3$

We are now ready to prove the initial theorem for prime numbers of the form $4n + 3$.

THEOREM 27.1. *A prime number p of the form $4n + 3$ cannot be represented as a sum of two squares.*

PROOF. If p is a prime of the form $4n + 3$ then by corollary 27.1 the value -1 is a non-residue. Let us suppose that we could write $a^2 + b^2 = p$ for integers a and b. Then:

$$a^2 + b^2 \equiv 0 \pmod{p} \quad \Rightarrow \quad a^2 \equiv (-1) \cdot b^2 \pmod{p}$$

But the above equation shows us that the multiplication of a non-residue (-1) by a residue (b^2) results in a residue (a^2), which is a contradiction with lemma 27.2. So the hypothesis that we can write $a^2 + b^2 = p$ is false. $\qquad\square$

If we look closely, we have not only proved the theorem for all primes of the form $4n + 3$, but also for every number **multiple** of a prime of the form $4n + 3$. For example, do not waste time looking for a sum of two squares that results in the number 7007, since it cannot exist.

To prove the other part of the theorem (prime numbers p of the form $4n + 1$) we will have to work a little more.

Norms

DEFINITION 27.3. *A natural number n is a **norm** if it can be written as the sum of two squares of natural numbers $n = a^2 + b^2$. In that case, we say that the values a and b are **bases** of n.*

Now we are going to prove the key part of the whole procedure, which surely eluded Euler's imagination for years: we will see that if an odd prime p divides a norm but not its bases (for example, 157 divides the norm $23^2 + 16^2 = 785$ since $785 = 5 \cdot 157$ but it does not divide either number 23 or 16) then that prime number p is also a norm ($157 = 6^2 + 11^2$).

For this proof we will use a demonstration technique that was discovered by Fermat and later used for many proofs of his results: the "infinite descent". This method, valid for proofs that involve natural numbers, consists of looking for an equality holding for a certain set of natural numbers and then to prove that it must be satisfied by a natural number that is smaller than any of the previous numbers of the set (a "descent"). Now we include this number in the initial set and we repeat the process infinitely many times ("infinite") to reach a conclusion (or a contradiction) since the set of natural numbers is bounded below by the number 1. The method does not work for real numbers, since there is no positive real number that is smaller than the rest.

PROPOSITION 27.2. *If an odd prime number p divides a norm but does not divide any of its bases, then p is a norm.*

PROOF. Let us suppose that an odd prime p divides a norm but does not divide any of its bases a_0, b_0:

$$(27.3) \qquad a_0^2 + b_0^2 = p \cdot f$$

We calculate a_1 and a_2 as the only integer numbers such that $a_1 = a_0 - m \cdot f$ ($|a_1| \leq f/2$, m integer) and $b_1 = b_0 - n \cdot f$ ($|b_1| \leq f/2$, n integer), i.e., these numbers are congruent with the original module f but with a minimum absolute value. It should be noted that, by hypothesis, since f does not divide either a_0 or b_0 then, for the previous inequalities, it follows that it does not divide either a_1 or b_1.

Substituting in equation (27.3):

$$(a_1 + mf)^2 + (b_1 + nf)^2 = p \cdot f \quad \Rightarrow \quad a_1^2 + 2a_1mf + m^2f^2 + b_1^2 + 2b_1nf + n^2f^2 = p \cdot f$$

$$(27.4) \qquad\qquad\qquad\qquad\qquad \Rightarrow \quad a_1^2 + b_1^2 = f' \cdot f$$

Since we have defined a_1 and b_1 such that $|a_1| \leq f/2$ and $|b_1| \leq f/2$ it holds that:

$$a_1^2 + b_1^2 \leq (f/2)^2 + (f/2)^2 = \frac{f}{2} \cdot f$$

Comparing this result with equation (27.4), we conclude that $f' \leq (f/2)$.

Now let us multiply both equations (27.3) and (27.4):

$$(27.5) \quad (a_0^2 + b_0^2)^2 \cdot (a_1^2 + b_1^2)^2 = p \cdot f^2 \cdot f' \quad \Rightarrow \quad (a_0a_1 + b_0b_1)^2 + (a_0b_1 - a_1b_0)^2 = p \cdot f^2 \cdot f'$$

As we can write:

$$(a_0a_1 + b_0b_1) = a_0 \cdot (a_0 + mf) + b_0 \cdot (b_0 + nf) = a_0^2 + b_0^2 + kf = pf + kf = a_2f$$
$$(a_0b_1 - a_1b_0) = a_0 \cdot (b_0 + nf) - b_0 \cdot (a_0 + mf) = b_2f$$

we can substitute in equation (27.5) to obtain:

$$(27.6) \qquad (a_2f)^2 + (b_2f)^2 = p \cdot f^2 \cdot f' \quad \Rightarrow \quad a_2^2 + b_2^2 = p \cdot f'$$

Let us recap what we have done so far: starting from equation (27.3) we have obtained equation (27.6) where all numbers are positive integers and $f' \leq f/2$. It is clear that $f' \neq 0$ since otherwise it would imply [in equation (27.4)] that $a_1 = b_1 = 0$ and, by definition of a_1 and b_1, that f divides a_0 and b_0, which is a contradiction to the initial hypothesis.

Therefore, we have equation (27.6) where $1 \leq f' \leq f/2$. If $f' = 1$ we would have already found a representation of p as a norm ($a_2^2 + b_2^2 = p$); otherwise, if $f' > 1$ we can repeat the process from (27.3) to (27.6) as many times as we need in order to get a representation of p as a norm (the process is finite since we are using natural numbers and in each step we calculate a new value that is, at most, half of the previous one: at some point necessarily we will get $f' = 1$). $\square$

Let us look at the example of equation $23^2 + 16^2 = 785 = 5 \cdot 157$. If we do the operations proposed in the proof ($-2 = 23 - 5 \cdot 5$ and $1 = 16 - 3 \cdot 5$) we would obtain the equation $2^2 + 1^2 = 5 \cdot 1$, and if we now multiply both of them we get:

$$(23^2 + 16^2) \cdot (2^2 + 1^2) = 157 \cdot 5^2 \cdot 1 \quad \Rightarrow$$

(27.7)
$$(23 \cdot 2 + 16 \cdot 1)^2 + (23 \cdot 1 - 16 \cdot 2) = 157 \cdot 5^2 \cdot 1$$

We calculate the following too:

$$(23 \cdot 2 + 16 \cdot 1) = 23 \cdot (23 - 5 \cdot 5) + 16 \cdot (16 - 3 \cdot 5) = 23^2 + 16^2 - 163 \cdot 5 = 157 \cdot 5 - 163 \cdot 5 = -6 \cdot 5$$

$$(23 \cdot 1 - 16 \cdot 2) = 23 \cdot (16 - 3 \cdot 5) - 16 \cdot (23 - 5 \cdot 5) = 11 \cdot 5$$

Substituting in equation (27.7) we get:

$$(6 \cdot 5)^2 + (11 \cdot 5)^2 = 157 \cdot 5^2 \cdot 1 \quad \Rightarrow \quad 6^2 + 11^2 = 157$$

which is what we want to obtain (157 as the sum of two squares). It should be noted that, in this case, the first iteration of the method has already been enough to find $f' = 1$ (a priori we knew that the value f' has to satisfy that $f' \leq 5/2$, so the possibility that $f' = 2$ was also feasible), but otherwise the method would be repeated until this value is achieved.

Proof of the general theorem for prime numbers of the form $4n + 1$

THEOREM 27.2. *Every odd prime p of the form $4n+1$ can be represented as a norm. Furthermore, there is only one way to do so.*

PROOF. If p is a prime of the form $4n+1$ then, by corollary 27.1, the value -1 is a residue, i.e., there exists a value x ($0 < x < p/2$) such that $x^2 \equiv -1 \pmod{p}$, which implies that $x^2 + 1 = k \cdot p$. But now, by proposition 27.2, we have a prime p that divides a norm but not any of its bases (both 1 and x are smaller than p), which implies that p is a norm, that is, it can be written as the sum of two squares.

It only remains to see that, in the latter case, p can be written as a sum of squares only in one way. Hence, suppose that we can write p twice as sum of squares: $p = a_1^2 + b_1^2$ and $p = a_2^2 + b_2^2$, where all numbers are positive integers. If we multiply both equations:

(27.8)
$$p^2 = (a_1^2 + b_1^2) \cdot (a_2^2 + b_2^2) = (a_1 a_2 \pm b_1 b_2)^2 + (a_1 b_2 \mp b_1 a_2)^2$$

where the equation is valid if we take both upper signs or both lower signs. Now, as the product of the factors $(a_1a_2 + b_1b_2)$ and $(a_1a_2 - b_1b_2)$:

$$(a_1a_2 + b_1b_2) \cdot (a_1a_2 - b_1b_2) = a_1^2a_2^2 - b_1^2b_2^2 = a_2^2 \cdot (a_1^2 + b_1^2) - b_1^2 \cdot (a_2^2 + b_2^2)$$

is divisible by p (since, by hypothesis, $p = a_1^2 + b_1^2$ and $p = a_2^2 + b_2^2$), at least one of the two factors $(a_1a_2 + b_1b_2)$ or $(a_1a_2 - b_1b_2)$ must also be divisible by p (a well-known property of any prime number). Consequently, we select in (27.8) the upper signs (if $a_1a_2 + b_1b_2$ is multiple of p) or the lower ones (if $a_1a_2 - b_1b_2$ is multiple of p).

In the first case ($a_1a_2 + b_1b_2$ is a multiple of p) it must be true that $a_1a_2 + b_1b_2 = p$, since all numbers are positive (and therefore $a_1a_2 + b_1b_2 \neq 0$) and, in addition, $a_1a_2 + b_1b_2 \leq p$ by equation (27.8). In the second case ($a_1a_2 - b_1b_2$ is a multiple of p) it must be true that $a_1a_2 - b_1b_2 = 0$, since if $|a_1a_2 - b_1b_2| \geq p$ we would obtain a contradiction when we try to satisfy the equation (27.8), $p^2 = (a_1a_2 - b_1b_2)^2 + (a_1b_2 + b_1a_2)^2$, because $(a_1b_2 + b_1a_2)^2 \geq 0$.

We have inferred that:

$$\begin{cases} a_1a_2 + b_1b_2 = p \quad \text{and} \quad a_1b_2 - b_1a_2 = 0 \\ \qquad\qquad \text{or} \\ a_1a_2 - b_1b_2 = 0 \quad \text{and} \quad a_1b_2 + b_1a_2 = p \end{cases}$$

In the first case, $a_1b_2 - b_1a_2 = 0$ is satisfied, which means that if $a_1 > a_2$ (respectively, $a_1 < a_2$), then $b_1 > b_2$ (respectively, $b_1 < b_2$) and we reach a contradiction because $a_1^2 + b_1^2 > a_2^2 + b_2^2$ (respectively, $a_1^2 + b_1^2 < a_2^2 + b_2^2$), which is impossible because we have begun our reasoning by assuming that $a_1^2 + b_1^2 = a_2^2 + b_2^2$. The only viable possibility is that $a_1 = a_2$ and $b_1 = b_2$.

Similarly, in the second case it is true that $a_1a_2 - b_1b_2 = 0$, which means that if $a_1 > b_2$ (respectively $a_1 < b_2$) then $b_1 > a_2$ (respectively $b_1 < a_2$) and we reach a contradiction again because $a_1^2 + b_1^2 > a_2^2 + b_2^2$ (respectively, $a_1^2 + b_1^2 < a_2^2 + b_2^2$), which is impossible by the same reasoning of the previous section. In this case, the only viable possibility is that $a_1 = b_2$ and $b_1 = a_2$.

In summary, in any of the possible cases we can conclude that both solutions are identical: two different ways of writing p as the sum of two squares do not exist. $\qquad\square$

FINAL REMARKS

As discussed in the introduction, in order to prove a theorem with a statement about sums of squares and prime numbers we have developed powerful mathematical tools (modular arithmetic, quadratic residuals, proof by "infinite descent", uniqueness of the solutions) that undoubtedly strengthen any mathematician when facing other types of problems, perhaps of greater practical use.

It is not strange then that Fermat's theorem (which we will see in the problem "Impossibility of sum of cubes"), a very simple statement, endured for centuries the attacks of the greatest mathematicians. It was not until 1995 that the English mathematician Andrew Wiles (who achieved the title of Sir thanks to his solution) managed to prove it completely.

Chapter 28

The fundamental theorem of calculus

(Newton – 1665)

PROBLEM

To prove that the integral of a function $f(x)$, understood as the area that is enclosed between the function and the X axis, can be calculated with the inverse function of the derivative.

HISTORY

The first attempts to calculate the tangent to a curve (at a point) are very old, since Greek mathematicians like Apollonius studied it, although without getting a general method. It was not until the 17th century that Kepler and Cavalieri, among others, began to use what we now know as the derivative function.

In addition, the interest of calculating areas and volumes enclosed by functions led to the theory of integrals, a completely separate discipline from the previous one (in spite of the insistence of many modern teachers of Mathematics to "define" the integral as the inverse of the derivative).

It was not until a few years later that two geniuses like Isaac Newton and Gotfreid Leibniz realized about the intimate relationship between the derivative and integral functions, which is now known as the Fundamental Theorem of Calculus because of its enormous importance.

Statue of Gotfreid Leibniz in Leipzig
Author: Ernest Hahnel
Photo: Ad Meskens (Wikimedia Commons)

Apparently, Newton was the first man to discover the relationship between both functions, since he obtained it in 1664 – 1666, a period when he made Mathematics research at home due to the closure of the University of Cambridge (where he was a student) caused by a plague epidemic. He wrote several treatises on the subject, but its publication was some years later, which fueled the controversy over whether it was Leibniz or himself who held the honorable title of inventor of the infinitesimal calculus.

Leibniz discovered the theorem and its implications in 1675, publishing two articles about the topic before Newton. Leibniz also put special interest in using an appropriate and easy-to-remember symbolism, with the inventions of the symbols $\int$ and dx used today.

In any case, the work of both served to bring Mathematics and its applications to a higher level in very diverse fields. The Fundamental Theorem of Calculus may actually be the most important achievement in mathematical history.

SOLUTION

Intuitively, the fundamental theorem of calculus can be explained with the help of figure 28.1. We have a continuous function $f(x)$ and we want to estimate of the area $A(x_0)$ that is enclosed between $f(x)$, the X axis and the vertical lines $x = 0$ and $x = x_0$ (that is, the integral function at point x_0).

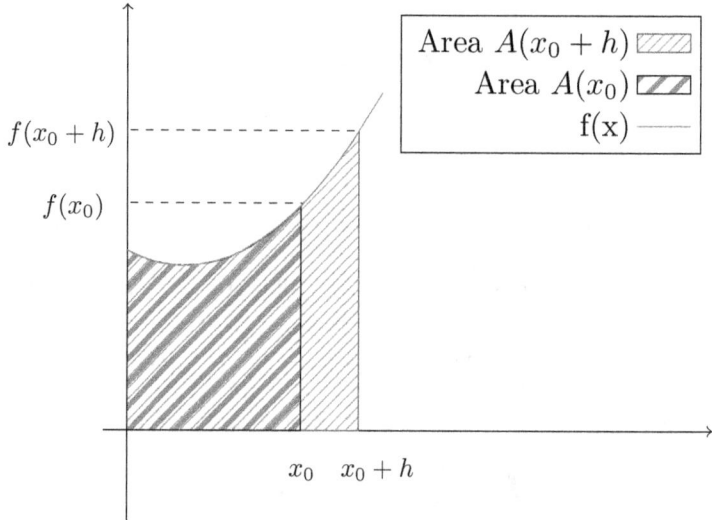

Figure 28.1

Let us also think of the value of $A(x_0 + h)$, the area contained between $f(x)$, the X axis and the vertical lines $x = 0$ and $x = x_0 + h$ (the integral function at point $x_0 + h$). If we assume that h is a very small value, $A(x_0 + h)$ has increased, with respect to $A(x_0)$, an amount that equals the value $f(x_0) \cdot h$ (the area of a rectangle of sides $f(x_0)$ and h, since we use the approximation $f(x_0) \approx f(x_0 + h)$). Then:

$$A(x_0 + h) - A(x_0) \approx f(x_0) \cdot h \quad \Rightarrow \quad f(x_0) \approx \frac{A(x_0 + h) - A(x_0)}{h}$$

Taking the limit ($h \to 0$), the value of $(A(x_0 + h) - A(x_0))/h$ tends to the derivative of $A(x)$ at point x_0, from which it follows that $f(x_0) = A'(x_0)$, that is, the function derived from the integral function is $f(x)$ or, what is the same, the derivative and integral functions are inverses between each other.

We are going to try to mathematically strengthen the previous idea. First we need a simple lemma.

LEMMA 28.1. *Suppose a continuous function $f(x)$ in the interval $[a, b]$ that reaches its minimum value m and its maximum value M in the aforementioned interval. Then:*

$$m \cdot (b - a) \le \int_a^b f(t) \cdot dt \le M \cdot (b - a)$$

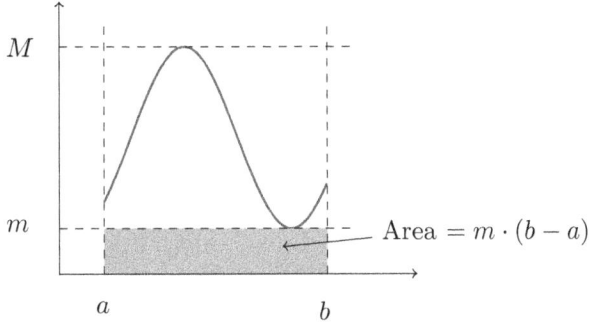

$$\text{Area} = m \cdot (b - a)$$

Figure 28.2

PROOF. The proof is obvious with the help of figure 28.2, since $m \cdot (ba)$ is the area of the rectangle of sides (ba) and m, which is less than or equal to (in the latter case, when $f(x)$ is a constant function) to the integral area of $f(x)$. Similarly, $M \cdot (b - a)$ is the area of the rectangle of sides $(b - a)$ and M, and this area is greater than or equal to the integral area. □

Now we are in good position to show a proof of the main theorem.

THEOREM 28.1. *(Fundamental Theorem of Calculus) Suppose a continuous function $f(x)$ in the interval $[a, b]$ and define its integral function $F(x)$ in $[a, b]$ as $F(x) = \int_a^x f(t) \cdot dt$. Then, $F'(x) = f(x)$ for all x in $[a, b]$.*

PROOF. Let c be a point of $[a, b]$. By definition, the derivative of the function $F(x)$ in c is:

(28.1) $$F'(c) = \lim_{h \to 0} \frac{F(c + h) - F(c)}{h}$$

We want to show that $F'(c) = f(c)$. To do this, first we will try to bound the value of $F(c+h) - F(c)$ and then we will take limits to its division by h, as indicated in equation (28.1).

By definition of function $F(x)$, the value of $F(c + h)$ is $\int_a^{c+h} f(t) \cdot dt$, while the value of $F(c)$ is $\int_a^c f(t) \cdot dt$, so we can write:

$$F(c + h) - F(c) = \int_a^{c+h} f(t) \cdot dt - \int_a^c f(t) \cdot dt = \int_c^{c+h} f(t) \cdot dt$$

Let m_c (resp., M_c) be the minimum value (resp., maximum) that the function $f(x)$ achieves in the interval $[c, c + h]$. Then, applying lemma 28.1:

$$m_c \cdot h \le F(c + h) - F(c) = \int_c^{c+h} f(t) \cdot dt \le M_c \cdot h$$

Substituting these inequalities in equation (28.1):

$$\lim_{h \to 0} m_c \leq F'(c) = \lim_{h \to 0} \frac{F(c+h) - F(c)}{h} \leq \lim_{h \to 0} M_c$$

But $f(x)$ is a continuous function in $x = c$, so:

$$\lim_{h \to 0} m_c = \lim_{h \to 0} M_c = f(c)$$

where we infer that $F'(c) = f(c)$. $\qquad\qquad\qquad\qquad\qquad\qquad\qquad\qquad\qquad\qquad$ $\square$

We have already completed the proof. However, the importance of the theorem is only highlighted when we state its most practical result: the Barrow rule to calculate the values of definite integrals.

COROLLARY 28.1. *(Barrow's rule) Let $f(x)$ be a continuous function in the interval $[a, b]$, and we define its integral function $F(x)$ in $[a, b]$ as $F(x) = \int_a^x f(t) \cdot dt$. Suppose we know a function $g(x)$, a primitive function of $f(x)$, such that $g'(x) = f(x)$ for all x in $[a, b]$. Then:*

$$\int_a^b f(t) \cdot dt = g(b) - g(a)$$

PROOF. Do not be confused about what we know: $F(x)$ is the function defined as the area enclosing function $f(x)$ and the X axis, but we do not have a "closed-form expression" for it (like $x^3 + \sin x$, for example); but we do have an explicit expression for function $g(x)$, because we have found it when we looked for a primitive of function $f(x)$.

What happens with these functions (an "unknown" one and a "known" one) is that both fulfill the property that its derivative is equal to $f(x)$: the first one, by the Fundamental Theorem of Calculus that we have proved previously; the second one, simply because we have searched for it when trying to satisfy this condition. That is, we have $F'(x) = g'(x) = f(x)$ for all x in $[a, b]$.

When two functions have the same derivative then they must be equal except for the sum of a constant value: therefore, it follows that $F(x) = g(x) + C$. To determine the value of constant C, let us evaluate both functions at the point $x = a$. By definition of $F(x)$:

$$F(a) = \int_a^a f(t) \cdot dt = 0$$

from where we infer that $F(a) = 0 = g(a) + C$, i.e., $C = -g(a)$ and $F(x) = g(x) - g(a)$.

If we now evaluate both functions at point $x = b$ we have that $F(b) = g(b) + C = g(b) - g(a)$. By definition of $F(x)$, it is satisfied that $F(b) = \int_a^b f(t) \cdot dt$, so we finally claim that:

$$\int_a^b f(t) \cdot dt = g(b) - g(a)$$

$\qquad\qquad\qquad\qquad\qquad\qquad\qquad\qquad\qquad\qquad\qquad\qquad\qquad\qquad\qquad\qquad$ $\square$

FINAL REMARKS

- Barrow's formula allows us to calculate definite integrals (provided that we are good enough to find a primitive function of $f(x)$). It should be noted that it does not matter what primitive we find, since the formula is true for any of them. Its name is due to the English mathematician Isaac Barrow (1630 – 1677), who had an important role in the development of integral calculus.

- Obviously, the calculation of areas is only the first utility of integrals. Its use to calculate infinite sums of actions defined by a function is much more important, which is always difficult to assimilate for the student who receives mainly the lesson to calculate areas and volumes.

- The Barrow formula is only a particular case of the Stokes Theorem, which relates "functions" in a domain (in Barrow formula, a one-dimension closed interval $[a, b]$) with its "primitive function" evaluated in the **boundary** of the domain (in Barrow formula, the values at the endpoints of the interval). The generalization applies to any dimension and to another type of "functions" that Stokes called differential forms. The understanding of Stokes theorem is outside the scope of this book.

Chapter 29

Power series of the logarithmic function

(Mercator – 1668)

PROBLEM

To calculate the natural logarithm of a number x using a power series.

HISTORY

The idea of the logarithm was born much earlier than the number e (as we will see in the problem "The number e", in the second part of this book), when the Scottish Baron John Napier (1550 – 1617) proposed it as a method to simplify calculations that involved very large numbers (demanded by the astronomical problems of his interest). The term "Neperian logarithm", derived from his surname, is still used nowadays.

Portrait of John Napier (unknown author)
Scottish National Gallery

Some years later, the German mathematician known as Nicolaus Mercator (1620 – 1687) (his real name was Kaufmann) discovered the relationship between the area under the hyperbola $f(x) = 1/(1 - x)$ (between lines $x = 0$ and $x = a$) and the logarithm of number a (as described by Napier). He devised a method that led him to the approach that today we would write as:

$$\ln(1 + x) = \frac{x}{1} - \frac{x^2}{2} + \frac{x^3}{3} + \cdots + (-1)^{n+1} \cdot \frac{x^n}{n} + \cdots \qquad \text{for } -1 < x < 1$$

published in his work "Logarithmotechnia" (London, 1668).

In the following simpler method, we will start from the fact (proven by Euler) that the function $\ln(x)$ is the inverse of the function e^x (which is not obvious).

SOLUTION

LEMMA 29.1. *The derivative function of $f(x) = \ln(x)$ is $f'(x) = 1/x$ for $x > 0$.*

PROOF. We can take advantage of the fact that the function $\ln(x)$ is the inverse of e^x to deduce its derivative.

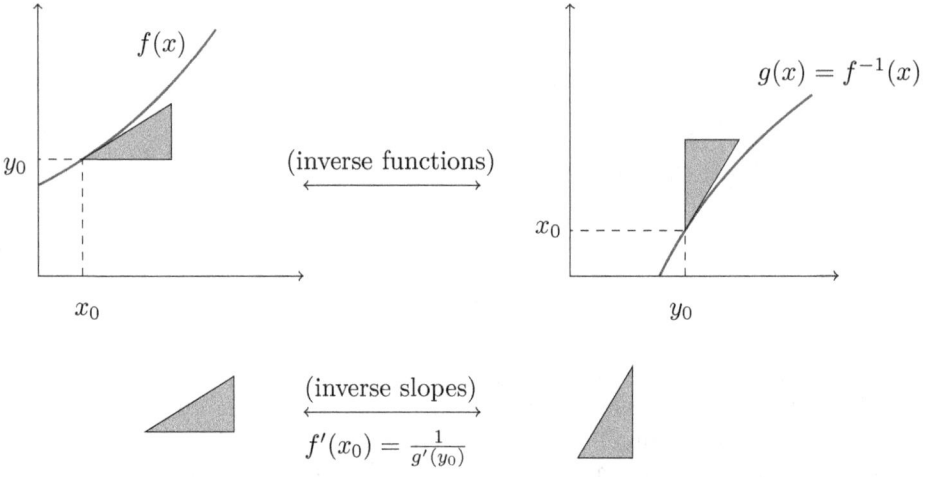

Figure 29.1

In calculus, the method of deriving a function of which we know the derivative of its inverse function is well known. If we want to calculate the value of the derivative function of $f(x)$ (whose inverse is $g(y) = f^{-1}(y)$) at point x_0 and we know the value of the derivative at point $y_0 = f^{-1}(x_0)$ of the function $f^{-1}(x)$, then we can do it in the following way:

$$(29.1) \qquad f'(x_0) = \frac{1}{g'(y_0)}$$

The validity of this formula can be seen in figure 29.1, where $f'(x_0)$ is, by definition, the slope of the tangent to $f(x)$ at point x_0, while $g'(y_0)$ is, by definition, the slope of the tangent to $g(y)$ at point y_0. By hypothesis, f and g are inverse functions (where $y_0 = f(x_0)$), so it follows that both slopes are inverses with respect to each other (formula 29.1).

In our case, the functions are $f(x) = \ln(x)$ and $g(y) = e^y$, so for all x_0 (where $\ln(x_0) = y_0$ or, which is the same, $x_0 = e^{y_0}$):

$$f'(x_0) = \frac{1}{g'(y_0)} = \frac{1}{e^{y_0}} = \frac{1}{x_0}$$

$\square$

LEMMA 29.2. *Let $h(x) = 1/(1+x)$. Then:*

$$(29.2) \qquad h(x) = 1 - x + x^2 - x^3 + \cdots + (-1)^{n-1} \cdot x^{n-1} + (-1)^n \cdot x^n \cdot h(x)$$

PROOF. It is trivial to see that:

$$(29.3) \qquad\qquad h(x) = 1 - x \cdot h(x)$$

since $h(x) = 1 - x \cdot (1/(1+x)) = (1 + x - x)/(1 + x) = h(x)$. But if we now apply (29.3) to the right-hand side of the same equation (29.3):

$$(29.4) \qquad h(x) = 1 - x \cdot (1 - x \cdot h(x)) \qquad \Rightarrow \qquad h(x) = 1 - x + x^2 \cdot h(x)$$

We apply again (29.3), now to the right part of (29.4), to find:

$$(29.5) \qquad h(x) = 1 - x + x^2 \cdot (1 - x \cdot h(x)) \qquad \Rightarrow \qquad h(x) = 1 - x + x^2 - x^3 \cdot h(x)$$

The repeated application of this procedure proves the statement of the lemma. $\qquad\square$

LEMMA 29.3. *Let n be a positive integer and let x be a real number such that $x > 0$. Then:*

$$\left| \int_0^x t^n \cdot \frac{1}{1+t} \cdot dt \right| < \frac{x^{n+1}}{n+1}$$

PROOF. The value of the integral is positive since its function is a multiplication of positive functions in the interval studied. To see the inequality: the function $f(t) = 1/(1+t)$ reaches its maximum in the interval $[0, x]$ ($x > 0$) at point $t = 0$, where its value is equal to $f(0) = 1$ (see figure 29.2). Therefore, the value of the integral will be smaller if we substitute the function $1/(1+t)$ (which is positive throughout the interval) by a constant function equal to its maximum value 1:

$$\left| \int_0^x t^n \cdot \frac{1}{1+t} \cdot dt \right| < \int_0^x t^n \cdot dt = \frac{x^{n+1}}{n+1}$$

$\qquad\square$

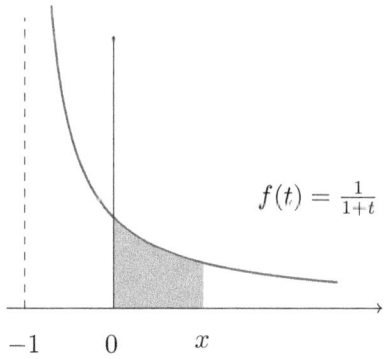

$$f(t) = \tfrac{1}{1+t}$$

$-1 \quad 0 \quad\quad x$

Case $x > 0$: Maximum at $t = 0$

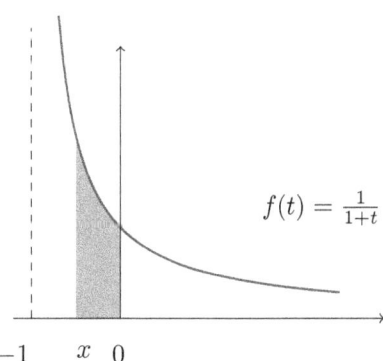

$$f(t) = \tfrac{1}{1+t}$$

$-1 \quad x \quad 0$

Case $x < 0$: Maximum at $t = x$

Figure 29.2

LEMMA 29.4. *Let n be a positive integer and let x be a real number such that $-1 < x < 0$. Then:*

$$\left| \int_0^x t^n \cdot \frac{1}{1+t} \cdot dt \right| < \frac{1}{1+x} \cdot \frac{x^{n+1}}{n+1}$$

PROOF. If $-1 < x < 0$, then the function $f(t) = 1/(1+t)$ reaches its maximum at point $t = x$ in the interval $[x, 0]$, ans its value is equal to $f(x) = 1/(1+x)$ (see figure 29.2). Therefore, the value of the integral will be smaller if we substitute the function $1/(1+t)$ (which is positive throughout the interval) by a constant function equal to its maximum value $1/(1+x)$:

$$\left| \int_0^x t^n \cdot \frac{1}{1+t} \cdot dt \right| < \frac{1}{1+x} \cdot \left| \int_0^x t^n \cdot dt \right| < \frac{1}{1+x} \cdot \frac{x^{n+1}}{n+1}$$

$\square$

After all these preliminary lemmas, we can prove the main theorem of the problem.

THEOREM 29.1. *For $-1 < x \leq 1$, the sum of terms $x - x^2/2 + x^3/3 + \cdots$ converges to the value $\ln(1+x)$.*

PROOF. If $x = 0$ the result is trivial. For the rest of values, we integrate both sides of expression 29.2 (we know how to integrate the function $1/(1+x)$ thanks to lemma 29.1 and the chain rule).

Let $x \neq 0$. Then:

$$\int_0^x \frac{1}{1+t} \cdot dt = \int_0^x (1 - t + t^2 - t^3 + \cdots + (-1)^{n-1} t^{n-1}) \cdot dt + R$$

$$\ln(1+x) = x - \frac{x^2}{2} + \frac{x^3}{3} - \frac{x^4}{4} + \cdots + (-1)^{n-1} \cdot \frac{x^n}{n} + R$$

where R is the error that is committed when we approximate up to the n-th term of the power series. If we can see that R tends to 0 when n tends to infinity, we would have proved the theorem.

In the case that $x > 0$, applying lemmas 29.2 and 29.3:

$$|R| = \left| \int_0^x t^n \cdot \frac{1}{1+t} \cdot dt \right| < \frac{x^{n+1}}{n+1}$$

If $0 < x \leq 1$, the right-hand side of the inequality tends to 0 when n tends to infinity.

When $-1 < x < 0$, applying lemmas 29.2 and 29.4:

$$|R| = \left| \int_0^x t^n \cdot \frac{1}{1+t} \cdot dt \right| < \frac{1}{1+x} \cdot \frac{x^{n+1}}{n+1}$$

If $-1 < x < 0$, the right-hand side of the inequality tends to 0 when n tends to infinity. $\square$

For example, the correct value of $\ln(1.2)$ is $0.1823215567...$. If we make the sum of the first six terms of the progression for $x = 0.2$, we will only have an approximation:

$$0.2 - \frac{(0.2)^2}{2} + \frac{(0.2)^3}{3} - \frac{(0.2)^4}{4} + \frac{(0.2)^5}{5} - \frac{(0.2)^6}{6} = 0.18232$$

Lemma 29.3 (the one to be used in this case, since $0 < x \leq 1$) assures us that the error of this approximation is less than $(0.2)^7/7 = 0.00000182$. This is true, since the real error is $0.1823215567... - 0.18232 = 0.0000015567...$.

FINAL REMARKS

Observation 1

With the theorem we can calculate the value of $\ln(y)$ for values such that $0 < y \leq 2$, but we must find an alternative method to calculate the logarithm for higher values. In order to do this, we replace $-x$ by x in the formula to get the two following power series:

(29.6)
$$\ln(1 + x) = x - \frac{1}{2}x^2 + \frac{1}{3}x^3 - \frac{1}{4}x^4 + \cdots \qquad -1 < x \leq 1$$

(29.7)
$$\ln(1 - x) = -x - \frac{1}{2}x^2 - \frac{1}{3}x^3 - \frac{1}{4}x^4 - \cdots \qquad -1 \leq x < 1$$

If we subtract (29.7) from (29.6), we obtain:

(29.8)
$$\ln\left(\frac{1+x}{1-x}\right) = 2 \cdot \left[x + \frac{x^3}{3} - \frac{x^5}{5} + \cdots\right] \qquad -1 < x < 1$$

The expression (29.8) allows us to calculate any value of a logarithm, since function $(1+x)/(1-x)$ takes any positive value when x takes all real values between -1 and 1. For example, to calculate $\ln(15)$ we use $x = 7/8$ in the formula:

$$\ln(15) - 2 \cdot \left[(7/8) + \frac{(7/8)^3}{3} - \frac{(7/8)^5}{5} + \cdots\right] = 2.7080502011...$$

Observation 2

However, the previous method converges very slowly for certain values of x (in fact, the closer its absolute value is to 1, the slower its convergence). In our example, with twelve terms in the series we would only get two correct decimals of $\ln(15)$.

Therefore, it is not necessary for us to be so restrictive and allow that in equation (29.8) the value of $(1 + x)/(1 - x)$ can be replaced by a quotient of two integers:

$$\frac{1+x}{1-x} = \frac{y_1}{y_2} \quad \Rightarrow \quad x = \frac{y_1 - y_2}{y_1 + y_2} \quad \Rightarrow$$

183

$$\Rightarrow \quad \ln(y_1) - \ln(y_2) = 2 \cdot \left[x + \frac{x^3}{3} - \frac{x^5}{5} + \cdots \right]$$

Now we should choose y_1 and y_2 suitably so that x is a value as close to 0 as possible, because the series would converge quickly. In our case, a possible solution would be to take $y_1 = 16$ and $y_2 = 15$, so $x = 1/31$:

$$\ln(16) - \ln(15) = 2 \cdot \left[(1/31) + \frac{(1/31)^3}{3} - \frac{(1/31)^5}{5} + \cdots \right] \quad \Rightarrow$$

$$\ln(15) = 4 \cdot \ln(2) - 2 \cdot \left[(1/31) + \frac{(1/31)^3}{3} - \frac{(1/31)^5}{5} + \cdots \right]$$

In this case it would be necessary to first know the correct value of $\ln(2)$, although once calculated only with two terms of the series we would get seven correct decimals.

Similar methods are used by calculators for logarithm calculations.

Chapter 30

Impossibility of sum of cubes

(Fermat – 1670)

PROBLEM

To prove that the sum of the cubes of two non-zero integers cannot be the cube of an integer, i.e., the equation $x^3 + y^3 = z^3$ has no solution for non-zero integers x, y and z.

HISTORY

The French lawyer Pierre de FERMAT (1601 – 1665), a great enthusiast of Mathematics, has passed to posterity for his contributions to Number Theory, a discipline that counts with many theorems of great importance stated by him. Interestingly enough, he proved almost none of them (he mainly limited himself to challenge others mathematicians of his era with some problems), but many of them were an inspiration to great mathematicians.

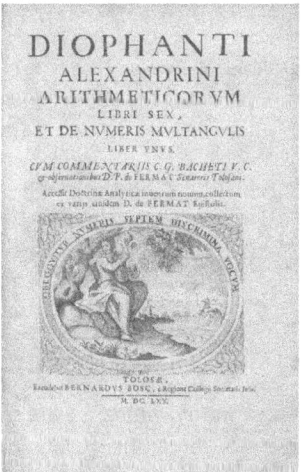

Cover of the book "Arithmetica " by Diophantus
(with comments from Fermat), 1670

The most famous theorem that he contributed was the one known as "Fermat's last theorem" or, simply, "Fermat's theorem". He neither proved it or proposed it to others mathematicians, but he wrote its statement in the margin of a book he was reading, as he often did. The work was a modern edition of "Arithmetica" from the Greek sage Diophantus and the commentary reads as follows (translated from French):

It is impossible to find a cube as the sum of two cubes, a fourth power as the sum of two fourth powers, or in general any power that is higher than two as the sum of two powers of the same degree. I have discovered an excellent proof for this fact but the margin of this book is too small to fit it in.

The problem, exposed to the world when the son of Fermat published a book in 1670 about the work of his father (then deceased), was studied by the most prestigious mathematicians but none of them managed to prove it until more than 300 years later.

One of those who devoted more time to the problem was the great Leonard EULER, who could not solve it in his general case; desperate after many years of attempts, the legend states that he did rummage through the belongings of Fermat in order to find some clue to find the solution. At the very least, Euler is credited with the proof for the case of the power $n = 3$, which is the one that concerns us, although the truth is that he made an error in it.

The extensive solution that we present below is based on Euler's proof, corrected for its presentation.

SOLUTION

Initial observations on the equation

Let us look at the equation:

$$(30.1) \qquad\qquad x^3 + y^3 = z^3$$

We are going to assume that we have an integer solution (x_0, y_0, z_0) such that $x_0 \neq 0$, $y_0 \neq 0$ and $z_0 \neq 0$. We will reach a contradiction in the reasoning that will come next, which will imply that the assumption is false and therefore that there is no solution satisfying these conditions.

Observation 1 We can assume, without loss of generality, that x_0, y_0, z_0 do **not** have a common divisor greater than 1, i.e., that they are coprime numbers.

PROOF. Suppose that x_0, y_0 have a prime common divisor m that is greater than 1 (the reasoning is identical if we assume that x_0, z_0 or y_0, z_0 are the numbers that share a divisor). Then we can write $x_0 = m \cdot x_1$, $y_0 = m \cdot y_1$; substituting in equation (30.1) we have that $m^3 x_1^3 + m^3 y_1^3 = z_0^3$, i.e., $z_0^3 \equiv 0 \pmod{m^3}$. But if z_0 is not a multiple of m then it is impossible for z_0^3 to be a multiple of m^3.

Therefore z_0 is a multiple of m, so we can write that $z_0 = m \cdot z_1$ and (x_1, y_1, z_1) is a solution of equation (30.1). We repeat the process as many times as necessary (in a finite number of steps, since we are working with integers) until the solutions do not have a common divisor. $\square$

From now on we will assume that (x_0, y_0, z_0) are coprime numbers. As a particular case, only one of them can be an even number; but, in fact, one of them must necessarily be an even number, since the equation (30.1) cannot be fulfilled if all three values are odd numbers (in that case, the result $x^3 + y^3$ would be an even number while z^3 would be an odd number)

Observation 2 We can assume that all values of the solution (x_0, y_0, z_0) are positive.

PROOF. In that case, if z_0 is negative, then $(-x_0, -y_0, -z_0)$ would be another solution where $-z_0$ is a positive integer. We assume then that we have a solution where z_0 is positive. Now both values x_0, y_0 cannot be negative if we want that the equation (30.1) to be fulfilled; if one of them, for example x_0, is positive and the other one, y_0, is negative then we would get $x_0^3 = (-y_0)^3 + z_0^3$, where all integers are positive. $\square$

In short, IF THERE IS a solution, then surely there is another one where all numbers are positive integers, pairwise coprimes and where one of the values, and only one, is an even integer.

Properties of the numbers of the form $a^2 + 3b^2$

In Euler's proof we will see the enormous importance of the integer numbers of the form $a^2 + 3b^2$, where a and b are positive integers. We will define these numbers in his honor as E-numbers and in this section we will study some of their properties, that will be fundamental for the proof of the theorem.

LEMMA 30.1. *The multiplication of two E-numbers is an E-number.*

For example, the multiplication of 79 ($= 2^2 + 3 \cdot 5^2$) and 49 ($= 1^2 + 3 \cdot 4^2$) results in 3871 ($= 58^2 + 3 \cdot 13^2$).

PROOF.
$$(a^2 + 3b^2) \cdot (c^2 + 3d^2) = a^2c^2 + 3a^2d^2 + 3b^2c^2 + 9b^2d^2 =$$
$$= (a^2c^2 - 6abcd + 9b^2d^2) + (3a^2d^2 + 6abcd + 3b^2c^2) =$$
$$= (ac - 3bd)^2 + 3 \cdot (ad + bc)^2$$

$\square$

In fact, we have not only proved the lemma, but we have found a formula for its easy calculation:

(30.2a) $$(a^2 + 3b^2) \cdot (c^2 + 3d^2) = (ac - 3bd)^2 + 3 \cdot (ad + bc)^2$$

(30.2b) $$(a^2 + 3b^2) \cdot (c^2 + 3d^2) = (ac + 3bd)^2 + 3 \cdot (ad - bc)^2$$

The second equation (30.2b) is obtained in a similar way (changing the sign of term $6abcd$ in the proof of lemma 30.1), and it leads us in the previous example to the equality $3871 = 62^2 + 3 \cdot 3^2$.

LEMMA 30.2. *The division of an E-number by a **prime** that is an E-number is an E-number.*

In the previous example, the division of 3871 (E-number) by 79 (prime that is an E-number) is another E-number, which is 49. The same statement could not be guaranteed if we divided by 3871 by 49 (because it is not a prime number) although, on this occasion, it would also be fulfilled.

PROOF. Let $(a^2 + 3b^2)$ be an E-number which is a multiple of a prime E-number $(p^2 + 3q^2)$: we want to prove that its quotient is an E-number. First, let us develop the expression $(pb - aq) \cdot (pb + aq)$:

$$(pb - aq) \cdot (pb + aq) = \cdots = b^2(p^2 + 3q^2) - q^2(a^2 + 3b^2) \quad \Rightarrow$$
$$\frac{(pb - aq) \cdot (pb + aq)}{p^2 + 3q^2} = b^2 - \frac{q^2(a^2 + 3b^2)}{p^2 + 3q^2}$$

The right-hand side of the equation is an integer (since we are assuming that $p^2 + 3q^2$ is a divisor of $a^2 + 3b^2$), so the left-hand side must also be an integer. That means that $p^2 + 3q^2$ divides the product $(pb - aq) \cdot (pb + aq)$; but, since $p^2 + 3q^2$ is a prime number, then it divides one term of the multiplication or the other one: i.e., $p^2 + 3q^2$ necessarily divides $pb + aq$ or $pb - aq$.

Let us suppose that $p^2 + 3q^2$ divides $pb + aq$. By equation (30.2a):

$$(p^2 + 3q^2) \cdot (a^2 + 3b^2) = (pa - 3bq)^2 + 3 \cdot (pb + aq)^2$$

Dividing by $p^2 + 3q^2$:

$$(30.3) \qquad a^2 + 3b^2 = \frac{(pa - 3bq)^2}{p^2 + 3q^2} + \frac{3 \cdot (pb + aq)^2}{p^2 + 3q^2}$$

The left-hand side of the equality is an integer, and so it is the second term of the sum in the right-hand side (we are assuming that $p^2 + 3q^2$ divides $pb + aq$), so it follows that the first term of the sum in the right-hand side must be an integer too; i.e., $p^2 + 3q^2$ divides $(pa - 3bq)^2$ and, since $p^2 + 3q^2$ is a prime number, this means that $p^2 + 3q^2$ divides $pa - 3bq$.

If we now divide equation (30.3) by $p^2 + 3q^2$ again:

$$\frac{a^2 + 3b^2}{p^2 + 3q^2} = \left[\frac{pa - 3bq}{p^2 + 3q^2}\right]^2 + 3 \cdot \left[\frac{pb + aq}{p^2 + 3q^2}\right]^2$$

We have seen that $p^2 + 3q^2$ divides both $pb + aq$ and $pa - 3bq$, so the above equation is the end of our proof.

The case where $p^2 + 3q^2$ divides $pb - aq$ is proved using equation (30.2b) instead of (30.2a). $\qquad \square$

LEMMA 30.3. *If an E-number is divisible by 2, then it is also divisible by 4. In addition, dividing the E-number by 4 results in another E-number.*

Since 4 is not a prime number, the second part of the lemma cannot be included in the previous lemma, so we should look for an alternative proof. Before that, I propose the following example: 148 is an E-number ($= 1^2 + 3 \cdot 7^2$) that is divisible by 2, so it is also a multiple of 4 ($148/4 = 37$) and its quotient is also an E-number ($37 = 5^2 + 3 \cdot 2^2$).

PROOF. Let $a^2 + 3b^2$ be an E-number. It is easy to see that if a and b have a different parity then $a^2 + 3b^2$ is odd, so it is not divisible by 2.

If a and b are even numbers then $a^2 + 3b^2$ is even too; but, in this case, $a^2 + 3b^2 = (2m)^2 + 3 \cdot (2n)^2 = 4(m^2 + 3n^2)$ and the lemma is true (the E-number is divisible by 4 and its quotient is an E-number).

Then only the case where a and b are both odd numbers is left ($a^2 + 3b^2$ is an even number then), which requires more effort. Since an odd number can be written as $4k \pm 1$, it is clear that either the sum or the subtraction of a and b is a multiple of 4. Let us suppose for example that 4 divides $a + b$ (the case where 4 divides $a - b$ is very similar and it will not be shown).

If 4 divides $a + b$, it also divides $a - 3b$ (since $a - 3b = (a + b) - 4b$). Furthermore, since $4 = 1^2 + 3 \cdot 1^2$, we can deduce from equation (30.2a) that:

$$4 \cdot (a^2 + 3b^2) = (a - 3b)^2 + 3 \cdot (b + a)^2$$

As we have seen before, 4 divides both $a + b$ and $a - 3b$ (and therefore 4^2 divides both $(a + b)^2$ and $(a - 3b)^2$), so the above equation leads to the fact that 4 has to divide $(a^2 + 3b^2)$, which is the first statement we want to prove. But if we divide the above equation by 4^2 then:

$$\frac{(a^2 + 3b^2)}{4} = \left[\frac{a - 3b}{4}\right]^2 + 3 \cdot \left[\frac{b + a}{4}\right]^2$$

All the expressions in brackets are integers, so the division of an E-number by 4 is also an E-number. □

LEMMA 30.4. *If an E-number has an odd divisor that is **NOT** an E-number, then its quotient also has an odd divisor that is **NOT** an E-number.*

For example, let us think of the E-number 325 ($= 5^2 + 3 \cdot 10^2$): one of its odd divisors is 5, which is not an E-number. Therefore the quotient of $325/5$, which is 65, must also have an odd divisor that is not an E-number (number 5 again).

PROOF. Let m be an E-number with an odd divisor n which is not an E-number:

(30.4) $$m = n \cdot (p_1 \cdot p_2 \cdots p_r)$$

where $p_1, p_2, ..., p_r$ is the decomposition into prime numbers of the quotient m/n. First, if any of the p_i numbers is 2, then by lemma 30.3 another p_j must be equal to 2 (when an E-number is divisible by 2, it is also divisible by 4). If we divide equation (30.4) by those two prime numbers we will get a new equation:

(30.5) $$m' = n \cdot (p'_1 \cdot p'_2 \cdots p'_{r-2})$$

where m' is an E-number (by lemma 30.3, the division of an E-number by 4 results in another E-number) and $p'_1, p'_2, ..., p'_{r-2}$ are the same primes as before but removing two of them (those equal to 2). This procedure can be done as many (finite) times as necessary until a prime number equal to 2 no longer appears in the equation.

Let us now suppose that in equation (30.5) we do not have any primes equal to 2. Now, if all numbers p'_i are E-numbers then we can apply lemma 30.2 and divide the equation (30.5) for each of them. For example, dividing by p'_1:

$$m'' = n \cdot (p'_2 \cdots p'_{r-2})$$

where m'' is an E-number (lemma 30.2 states that the division of an E-number by a prime E-number is another E-number). Repeating the process $r - 2$ times we would reach the conclusion that:

$$m''''' = n$$

where in the left-hand side there is an E-number but in the right-hand side there a number that is not, which is a contradiction. Therefore the assumption that all numbers p_i are E-numbers was incorrect: at least one is not an E-number, which proves the lemma. $\square$

Finally, we announce a curious property of E-numbers.

PROPOSITION 30.1. *Let $a^2 + 3b^2$ be an E-number such that $(a, b) = 1$ (i.e., $gcd(a, b) = 1$). Then all its odd divisors (greater than 1) are also E-numbers.*

Notice that in our example of lemma 30.4 it is not true that $(a, b) = 1$, since $(5, 10) \neq 1$. The proposition that we are going to prove now states that we cannot find an example for lemma 30.4 if $(a, b) = 1$.

PROOF. Suppose that there exists an odd integer number x, greater than 1, with both the properties of being the smallest integer that is NOT an E-number and of being a divisor of an E-number whose coefficients are coprime. We will see below that if that number exists, then we could find another one which is smaller, positive, odd and greater than 1 and fulfilling the same properties (NOT an E-number and dividing an E-number whose coefficients are coprime).

But this is a contradiction (if x is the smaller natural number with this property, there cannot be another one with this property), so the assumption is incorrect and a natural number with this property cannot exist, which is what we want to prove.

So let x be a natural number, greater than 1, that is not an E-number and that it is a divisor of an E-number that we write as $a^2 + 3b^2$, where $(a, b) = 1$. First, we divide the values of a and b by x in such a way that the absolute value of the remainder is smaller than $x/2$:

$$(30.6) \qquad \begin{cases} a = mx + c & |c| < x/2 \\ b = nx + d & |d| < x/2 \end{cases}$$

Notice that, since x is odd, we can assure that the remainder of the division is smaller than $x/2$, which is essential for the subsequent reasoning (in case that x is even this property is not true: for example, dividing 20 by 8 we get $20 = 2 \cdot 8 + 4$ or $20 = 3 \cdot 8 - 4$, i.e., the remainder is smaller **or equal** to $x/2$).

We can write then:

$$a^2 + 3b^2 = (mx + c)^2 + 3 \cdot (nx + d)^2 = x \cdot f + (c^2 + 3d^2)$$

where f is an integer. Since x is a divisor of $a^2 + 3b^2$, from the above equation we can infer that x has to be a divisor of $c^2 + 3d^2$ too. In addition, since $(a, b) = 1$ we can assure that $c^2 + 3d^2 \neq 0$ (if $c^2 + 3d^2 = 0$ then $c = 0$ and $d = 0$, which would mean, in equation (30.6), that x would be a common divisor of a and b, which contradicts the hypothesis that they are coprime).

We can now write:

$$(30.7) \qquad c^2 + 3d^2 = x \cdot g$$

where g is an integer. The value of $c^2 + 3d^2$ is bounded, since we have $c^2 + 3d^2 < (x/2)^2 + 3\cdot(x/2)^2 = x^2$ (using inequalities in (30.7)). If we compare this inequality with equation (30.8) we conclude that $g < x$.

Since x is NOT an E-number (by hypothesis), lemma 30.4 ensures that any odd divisor of g (let us say it is x') is not an E-number either. This divisor x' would be greater than 1 (since $c^2 + 3d^2 \neq 0$ and 1 is an E-number) but smaller than x (since a divisor of g is smaller or equal to g itself, while we have seen that g is **strictly** smaller than x – here is the important point of the whole proof).

That is, we have found an odd number x' that is smaller than x but greater than 1, which is not an E-number; and such that it is a divisor of an E-number ($c^2 + 3d^2$). As we have already mentioned, this is a contradiction with the fact that x is the smallest natural number that fulfills these properties. The conclusion is that there is no number that meets these conditions. □

Representations of E-numbers

Although we have already seen some properties of E-numbers, we have to continue studying them. Suppose that a number n is an E-number: for each pair of positive values (a, b) such that $n = a^2 + 3b^2$ we say that we have a **representation** of n.

For example, the E-number 76 has three different representations:

$$\begin{cases} 76 = 1^2 + 3 \cdot 5^2 \\ 76 = 7^2 + 3 \cdot 3^2 \\ 76 = 8^2 + 3 \cdot 2^2 \end{cases}$$

The first two representations are using coprime numbers as coefficients $((1,5) = (7,3) = 1)$, but the third one does not meet this condition since $(8,2) = 2$. The first two representations are called **primitives** and they will be the only ones that we will consider from now on.

LEMMA 30.5. *A prime E-number only has one representation, and this representation is always primitive except for $p = 3$.*

The proof is very similar to the one in theorem 27.2 of the problem "Sum of squares". We have to be careful just to adapt the reasoning for numbers of the form $a^2 + b^2$ instead of numbers of the form $a^2 + 3b^2$. The reader will have no problem to adapt the proof for this case.

As an example you can verify that the only representation of 103 is $10^2 + 3 \cdot 1^2$. It is trivial to observe that the representations of prime numbers have to be primitive except for $p = 3$ (in this case, $3 = 0^2 + 3 \cdot 1^2$). Finally, we must bear in mind that there are primes that are NOT E-numbers, for example the number 101.

LEMMA 30.6. *Let x be an E-number that can be written as $x = a^2 + 3b^2$. Suppose that x is the result of multiplying a prime E-number r (whose only representation is $r = p^2 + 3q^2$) and another E-number s of which we know a representation $s = m^2 + 3n^2$. Then:*

(30.8)
$$\begin{cases} a = pm + 3qn \qquad b = pn - qm \\ \qquad\qquad or \\ a = pm - 3qn \qquad b = pn + qm \end{cases}$$

In short, for each known representation of s we can have at most two representations of x. For example, let us take the number $1729 = 19 \cdot 91$, where 19 is a prime number: the lemma states that for each representation of 91 we can find, by applying equations (30.8), two representations of 1729 at most. As 91 has two possible representations (we will see later why), $91 = 4^2 + 3 \cdot 5^2 = 8^2 + 3 \cdot 3^2$, equations (30.8) can give us a maximum of four representations of 1729 (if they are different). Indeed, in this case we obtained four different representations of 1729:

$$1729 = 1^2 + 3 \cdot 24^2 = 23^2 + 3 \cdot 20^2 = 31^2 + 3 \cdot 16^2 = 41^2 + 3 \cdot 4^2$$

The importance of the lemma (assuming that we trust that there are only two representations of 91) is that it assures us that another representation of 1729 other than the previous ones cannot exist. Let us see the proof, in which we are always going to choose positive values for representations (and also in the rest of this section).

PROOF. Let us use lemma 30.2, which states that $p^2 + 3q^2$ has to divide $pb + aq$ or $pb - aq$. Suppose you divide it by $pb + aq$; then, we saw that $p^2 + 3q^2$ divides $pa - 3bq$ and:

$$\begin{cases} pb + aq = m \cdot (p^2 + 3q^2) \\ pa - 3bq = n \cdot (p^2 + 3q^2) \end{cases}$$

If we solve this system of linear equations considering a and b as unknowns values, we will get that $a = pm + 3qn$ and $b = pn - qm$.

Otherwise, if we assume that $p^2 + 3q^2$ divides $pb - aq$ then the same reasoning would lead us to $a = pm - 3qn$ and $b = pn + qm$. There are no other possible solutions. $\square$

LEMMA 30.7. *Let n be an odd E-number that can be written as a multiplication of r different prime E-numbers (and all of them different from 3). Then, n has exactly 2^{r-1} different representations and all of them are primitive.*

We have already seen that a prime number (where $r = 1$) has one different representation ($2^{1-1} = 1$), which is also primitive. In other examples we have assured without proof that $91 = 13 \cdot 7$ (where $r = 2$) has two different representations ($2^{2-1} = 2$), all of them primitives, and that $1729 = 13 \cdot 7 \cdot 19$ (where $r = 3$) has four different representations ($2^{3-1} = 4$), also all primitives. Let us prove that this is not a coincidence but a mathematical truth.

The case when a prime is equal to 3 is different: for example, $21 = 3 \cdot 7$ has only one representation instead of two ($21 = 3^2 + 3 \cdot 2^2$) and $273 = 3 \cdot 7 \cdot 13$ has only two representations instead of four ($273 = 9^2 + 3 \cdot 8^2$ and $273 = 15^2 + 3 \cdot 4^2$). We let the reader prove that if the number n can be written as $3^t \cdot p_1 \cdot p_2 \cdots p_r$ (where all p_i are different from 3, different from each other and odd), then it has 2^{r-1} representations, all different and primitive (i.e., multiples of 3 in its decomposition into primes do not affect the calculation of representations of a number).

PROOF. Let $n = p_1 \cdot p_2 \cdots p_r$, where p_i is an odd prime E-number (different from 3) and where $p_i \neq p_j$ if $i \neq j$. By lemma 30.5, p_1 has only one representation. Therefore, by lemma 30.6, $p_1 \cdot p_2$ has two representations; hence, by the same lemma, $(p_1 \cdot p_2) \cdot p_3$ has four representations and so on, until we multiply all r prime E-numbers and then n has exactly 2^{r-1} representations. What we want to prove now is that all of them are different and primitive.

If two representations of n were equal to $a^2 + 3b^2$ (for example), that would mean that in the last step of the previous procedure (when we multiply $p_1 \cdot p_2 \cdots p_{r-1}$ by p_r) we could apply lemma 30.6 to deduce the following equations:

$$\begin{cases} a = pm_1 \pm 3qn_1 & b = pn_1 \mp qm_1 \\ \qquad\qquad y \\ a = pm_2 \pm 3qn_2 & b = pn_2 \mp qm_2 \end{cases}$$

where $m_1^2 + 3n_1^2 = m_2^2 + 3n_2^2$ are two representations of $p_1 \cdot p_2 \cdots p_{r-1}$ and $p^2 + 3q^2$ is the only representation of p_r. But in that case it follows that (applying the previous equations):

$$\begin{cases} pm_1 \pm 3qn_1 = pm_2 \pm 3qn_2 \\ pn_1 \mp qm_1 = pn_2 \mp qm_2 \end{cases} \Rightarrow \begin{array}{l} p(m_1 - m_2) = \pm 3q(n_1 - n_2) \\ q(m_1 - m_2) = \pm p(n_1 - n_2) \end{array}$$

If $m_1 \neq m_2$ (and therefore $n_1 \neq n_2$) then dividing both equations (it can be done provided that p and q are different from 0, so this is where we exclude the case where $p^2 + 3q^2 = 3$) we get $p/q = \pm 3q/p$, i.e., $p^2 = \pm 3q^2$, which is a contradiction with the fact that $p^2 + 3q^2$ is a prime number. Consequently, we deduce that $m_1 = m_2$ (and therefore $n_1 = n_2$).

In short, two equal representations of $p_1 \cdot p_2 \cdots p_r$ imply two equal representations of $p_1 \cdot p_2 \cdots p_{r-1}$. Repeating the procedure we would conclude that there are two equal representations of $p_1 \cdot p_2$, which is impossible by applying again the equations of lemma 30.6.

It only remains to prove that all representations are primitive. Let us suppose that at least one of them is NOT primitive, i.e., $n = a^2 + 3b^2$ where $(a, b) = d > 1$. In that case, $a^2 + 3b^2 = (d \cdot a')^2 + 3 \cdot (d \cdot b')^2 = d^2 \cdot [(a')^2 + 3 \cdot (b')^2]$ and therefore:

$$d^2 \cdot [(a')^2 + 3 \cdot (b')^2] = p_1 \cdot p_2 \cdots p_r$$

Now from the above equation we can infer that one number p_i has to divide d: let us suppose that it is p_1, i.e., $d = k \cdot p_1$ where k is an integer. Substituting:

$$(k \cdot p_1)^2 \cdot [(a')^2 + 3 \cdot (b')^2] = p_1 \cdot p_2 \cdots p_r \quad \Rightarrow \quad k^2 \cdot p_1 \cdot [(a')^2 + 3 \cdot (b')^2] = p_2 \cdots p_r$$

We deduce that p_1 divides $p_2 \cdots p_r$ and, as it is a prime number, p_1 divides one of the factors. But a prime number divide another prime number if only if they are equal, which is a contradiction with our hypothesis. $\square$

LEMMA 30.8. *Let n be an odd E number such that $n = t^r$, where t is an odd prime E-number (other than 3) and $r \geq 1$. Then n only has one primitive representation.*

For example, 49 has two representations but only one ($1^2 + 3 \cdot 4^2$) is a primitive one (the other one is $7^2 + 3 \cdot 0^2$); 343 also has two representations but only one ($10^2 + 3 \cdot 9^2$) is primitive (the other one is $14^2 + 3 \cdot 7^2$). The proof is not simple.

PROOF. Let $t = p^2 + 3 \cdot q^2$ (where $p \neq 0 \neq q$) and let $a_n^2 + 3 \cdot b_n^2$ be a representation of t^n ($p = a_1$, $q = b_1$). From equations (30.8) we can write the recursive equations:

$$\begin{cases} a_{n+1} = a_n p \mp 3b_n q \\ b_{n+1} = a_n q \pm b_n p \end{cases}$$

193

We say that the first choice of signs ($a_{n+1} = a_n p - 3 b_n q$, $b_{n+1} = a_n q + b_n p$) to calculate (a_{n+1}, b_{n+1}) is **principal**, while the second one is **secondary**. We are going to prove that:

- When we are calculating a representation of t, if we choose at any point the secondary option of signs to calculate (a_{n+1}, b_{n+1}) from (a_n, b_n), then we will obtain a non-primitive representation.
- If we always choose the principal option of signs instead, then we will obtain a primitive representation.

To prove the first statement, we have to observe that the calculation of representations is transitive: for example, if we choose the secondary option from $p^2 + 3q^2$ to calculate (a_2, b_2) and then the principal option to calculate (a_3, b_3), the resulting representation of t^3 would be the same as if we had chosen the principal option in the first step and the secondary option in the second step (we leave the details for the reader).

Therefore, if we use the secondary option at any time we can suppose that it was used in the first step. But then:

$$\begin{cases} a_2 = p \cdot p + 3q \cdot q \\ b_2 = p \cdot q - q \cdot p \end{cases}$$

and, as $b_2 = 0$, we have that $a_3 = a_2 \cdot p \pm 0$ and $b_3 = a_2 \cdot q \mp 0$, which implies that (a_3, b_3) > 1 and, applying equations (30.8), that (a_j, b_j) > 1 for all $j > 3$.

To prove the second statement, let us suppose that (a_r, b_r) $= d > 1$. This means that there exist a'_r, b'_r such that $a_r = d \cdot a'_r$, $b_r = d \cdot b'_r$ where (a'_r, b'_r) $= 1$. Substituting in the principal option of signs in the recursive equations:

(30.9)
$$\begin{cases} d \cdot a'_r = a_{r-1} \cdot p - 3 b_{r-1} \cdot q \\ d \cdot b'_r = a_{r-1} \cdot q + b_{r-1} \cdot p \end{cases}$$

Multiplying the first equation by p, the second equation by $3q$ and then adding both results:

(30.10)
$$d \cdot (p a'_r + 3 q b'_r) = a_{r-1} \cdot t$$

while multiplying the first equation by $-q$, the second equation by p and then adding both results:

(30.11)
$$d \cdot (p b'_r - q a'_r) = b_{r-1} \cdot t$$

From equations (30.10) and (30.11) we deduce that d divides $a_{r-1} \cdot t$ and $b_{r-1} \cdot t$. We have three possible options, considering that t is prime:

- d divides a_{r-1} or d divides b_{r-1}. Depending on the case, we deduce in equations (30.9) that, in fact, it divides both. Therefore, d divides $gcd(a_{r-1}, b_{r-1})$, i.e., (a_{r-1}, b_{r-1}) > 1.
- $d = k \cdot t$, where $k > 1$ and k divides both a_{r-1} and b_{r-1}. Again, d divides $gcd(a_{r-1}, b_{r-1})$, i.e., (a_{r-1}, b_{r-1}) > 1.

194

- $d = t$. In this case, equations (30.10) and (30.11) remain as:

$$\begin{cases} p \cdot a'_r + 3q \cdot b'_r = a_{r-1} \\ p \cdot b'_r - q \cdot a'_r = b_{r-1} \end{cases}$$

where we infer that a_{r-1}, b_{r-1} were calculated with the secondary option of signs from a'_r, b'_r (which is a representation of t^{r-1} since $d = t$ and therefore $a_r^2 + 3_r^2 = t^2 \cdot \left[(a'_r)^2 + 3(b'_r)^2 \right]$. But this is a contradiction since we have assumed that we have always chosen the principal option of signs.

In summary, only the cases that imply $(a_{r-1}, b_{r-1}) > 1$ survive. But now, repeating this argument as many times as needed, we would also deduce that $(a_2, b_2) > 1$, which is not true ($a_2 = p^2 - 3q^2$, $b_2 = 2p \cdot q$ but they do not have a common divisor if we recall the conditions of p and q in the statement). $\square$

From all the previous lemmas, we can present the following corollaries whose proofs are left to the reader.

COROLLARY 30.1. *Let n be an E-number such that $n = 3^t \cdot p_1 \cdot p_2 \cdots p_r$ (all different odd prime E-numbers and different from 3). Then n has 2^{r-1} representations, all different and primitive.*

COROLLARY 30.2. *Let n be an odd E-number of the form $n = p_1^{a_1} \cdot p_2^{a_2} \cdots p_r^{a_r}$, where p_i are different odd prime E-numbers and different from 3. Then n has exactly 2^{r-1} primitive representations.*

Example: $1609699 = 7^3 \cdot 13 \cdot 19^2$ has twelve representations, but only four of them are primitive ($124^2 + 3 \cdot 729^2$, $568^2 + 3 \cdot 655^2$, $1096^2 + 3 \cdot 369^2$ and $1268^2 + 3 \cdot 25^2$). In short, it does not matter what exponent each prime E-number has (by lemma 30.8); what is important is the number of different prime E-numbers in the decomposition of n (by lemma 30.7).

Finally, the key proposition for the resolution of the original problem is:

PROPOSITION 30.2. *Let $a^2 + 3b^2$ be an odd E-number that is not a multiple of 3 but it is a cube of an integer, with a, b positive integers such that $(a, b) = 1$. Then there exist integers t, w such that $a^2 + 3b^2 = \left(t^2 + 3 \cdot w^2 \right)^3$, where $a = t \cdot (t^2 - 9w^2)$ and $b = 3w \cdot (t^2 - w^2)$.*

PROOF. Let $n = a^2 + 3b^2$ be a number and $n = p_1^{a_1} \cdot p_2^{a_2} \cdots p_r^{a_r}$ be its decomposition into primes. Since n is odd and $(a, b) = 1$ then, by proposition 30.1, all its divisors are E-numbers. As it is a cube, the exponents $a_1, ..., a_r$ have to be multiples of 3 and we can write $n = \left[p_1^{b_1} \cdot p_2^{b_2} \cdots p_r^{b_r} \right]^3$, where $b_i = a_i/3$ are integers. Let m be the number $m = p_1^{b_1} \cdot p_2^{b_2} \cdots p_r^{b_r}$.

By corollary 30.2, both m and n have a total of 2^{r-1} different primitive representations (since neither of them is a multiple of 3 and all numbers p_i are odd, prime and E-numbers). Let $c_i^2 + 3d_i^2$ be a primitive representation of m and let us calculate from it the representation of m that results from applying the principal option of signs in equations (30.8) twice:

$$(c_i^2 + 3d_i^2) \cdot (c_i^2 + 3d_i^2) = (c_i^2 - 3d_i^2)^2 + 3 \cdot (2c_i d_i)^2$$

and

$$\left[(c_i^2 - 3d_i^2)^2 + 3 \cdot (2c_i d_i)^2 \right] \cdot (c_i^2 + 3d_i^2) = \left[c_i \cdot (c_i^2 - 9d_i^2) \right]^2 + 3 \cdot \left[3d_i \cdot (c_i^2 - d_i^2) \right]^2$$

So for each primitive representation of m we can find one and only one primitive representation of n (primitive and unique because of lemma 30.8). Since there is no other way to get a representation of n (as we saw in lemma 30.8, choosing the secondary option of signs in any step would contradict the fact that $(a, b) = 1$), this means that all representations of n are different.

I.e., for each primitive representation of n (including the target $a^2 + 3b^2$) there is a representation of m, $c_i^2 + 3d_i^2$, such that $a^2 + 3b^2 = (c_i^2 + 3d_i^2)^3$, $a = c_i \cdot (c_i^2 - 9d_i^2)$ and $b = 3d_i \cdot (c_i^2 - d_i^2)$. It is only necessary to rename $t = c_i$ and $w = d_i$. $\qquad \square$

As an example, let us take numbers $m = 7 \cdot 19^2 = 2527$ and $n = (7 \cdot 19^2)^3 = 16136737183$. Both numbers only have two primitive representations ($2^2 + 3 \cdot 29^2$ and $50^2 + 3 \cdot 3^2$ for m; $15130^2 + 3 \cdot 72819^2$ and $120950^2 + 3 \cdot 22419^2$ for n), and each representation of n can be written as the statement from coefficients of the representation of m from which it comes ($15130 = 2 \cdot (2^2 - 9 \cdot 29^2)$ and $72819 = 3 \cdot 29 \cdot (2^2 - 29^2)$ in the first case; $120950 = 50 \cdot (50^2 - 9 \cdot 3^2)$ and $22419 = 3 \cdot 3 \cdot (50^2 - 3^2)$ in the second one).

Proof of Fermat's Theorem for $n = 3$

After our exhaustive study of the properties of E-numbers, we are now ready to prove Fermat's Theorem as proposed by Euler.

To do this, let us suppose that there is a solution of the equation (30.1), i.e., $x_0^3 + y_0^3 = z_0^3$, and we will get a contradiction. As we saw in the first part, we can assume that the values x_0, y_0, z_0 are all positive and prime numbers, and that only one of them is even.

PROPOSITION 30.3. *If we define the numbers p and q as:*

- *a) If z_0 is even and, assuming for example that $x_0 > y_0$, we define $p = (x_0 + y_0)/2$, $q = (x_0 - y_0)/2$ (p and q are integers since if z_0 is even then x_0, y_0 are odd and, therefore, their addition and subtraction are even).*
- *b) If for example x_0 is even (the case y_0 even is symmetric), we define $p = (z_0 - y_0)/2$, $q = (z_0 + y_0)/2$ (p and q are integers by a similar reasoning to the one used in the previous case).*

then p and q are both positive integers, of different parity, coprime and such that as $p \cdot (p^2 + 3q^2)$ is a cube.

PROOF. It is clear that p and q are positive, since we have assumed in (a) that $x_0 > y_0$ (otherwise, we would have defined $q = (y_0 - x_0)/2$) and we know that $z_0 > y_0$ in (b) (because all values (x_0, y_0, z_0) are positive and $x_0^3 + y_0^3 = z_0^3$).

Furthermore, if p and q have a common divisor that is greater than 1, then x_0 and y_0 also have a common divisor in case (a) (since $x_0 = p + q$ and $y_0 = p - q$), which is a contradiction since we have assumed that these two values are coprime. Case (b) is analogous with y_0 and z_0.

To prove that they have different parity, we know that in case (a) x_0 is odd and $x_0 = p + q$, so necessarily one must be even and the other one odd (the reasoning is identical for case (b)).

Finally, it is true in case (a) that $2p \cdot (p^2 + 3q^2) = z_0^3$ (so it is a cube), since $z_0^3 = x_0^3 + y_0^3 = (x_0 + y_0) \cdot (x_0^2 - x_0 y_0 + y_0^2)$, that is:

$$z_0^3 = (2p) \cdot ((p + q)^2 - (p + q) \cdot (p - q) + (p - q)^2) = 2p \cdot (p^2 + 3q^2)$$

while it is satisfied in case (b) that $2p \cdot (p^2 + 3q^2) = x_0^3$ (a cube) since $x_0^3 = z_0^3 - y_0^3 = (z_0 - y_0) \cdot (z_0^2 + z_0 y_0 + y_0^2)$, that is:

$$x_0^3 = (2p) \cdot \left((p+q)^2 + (p+q) \cdot (q-p) + (p-q)^2\right) = 2p \cdot (p^2 + 3q^2)$$

$\square$

PROPOSITION 30.4. *Let p and q be numbers with the properties described in the previous proposition. Then, the greatest common divisor of $2p$ and $p^2 + 3q^2$ can only be 1 or 3.*

PROOF. Let k be a prime that divides both $2p$ and $p^2 + 3q^2$.

We can first prove that k is not equal to 2, since $p^2 + 3q^2$ is odd (in proposition 30.3 we saw that p and q are of different parity, so their squares also have different parity; multiplying a number by 3 does not change its parity, so p^2 and $3q^2$ have different parity and their sum is then an odd number).

Furthermore, let us suppose that k is greater than 3. Since k is a divisor of $2p$, then k is a divisor of p too (this is true because k is a prime number) and therefore it is a divisor of p^2; also, k is a divisor of $p^2 + 3q^2$ (remind that k it is a common divisor of $2p$ and $p^2 + 3q^2$ by hypothesis) and we have seen that it is a divisor of p^2, so it follows then that it is a divisor of the subtraction of both, that is, $3q^2$. But now, since k is greater than 3 and it is a prime number, then it is a divisor of q^2 and also a divisor of q. So, in this paragraph we have seen that k divides both p and q, but this is a contradiction with proposition 30.3, which assured us that p and q are coprime numbers. Therefore, k cannot be greater than 3.

In short, as a common divisor of $2p$ and $p^2 + 3q^2$ cannot be neither 2 nor contain any prime greater than 3, we only have left the possibility that the maximum common divisor is 1 or 3. $\square$

We have to see now that these two cases (maximum common divisor equal to 1 or equal to 3) lead us to a contradiction too. First we will show it for the case of maximum divisor equal to 1, thanks to the properties of the E-numbers that we worked so hard to deduce.

PROPOSITION 30.5. *Let (x_0, y_0, z_0) be a solution to equation (30.1) with positive and coprime values. Let p and q also be the numbers calculated as in proposition 30.3. If the greatest common divisor of $2p$ and $p^2 + 3q^2$ is 1, then there is a new solution (x_1, y_1, z_1) to equation (30.1), also with positive coprime values, which also satisfies the property that the product of its values $x_1^3 \cdot y_1^3 \cdot z_1^3$ is **smaller** than the product of the original values $x_0^3 \cdot y_0^3 \cdot z_0^3$.*

PROOF. First, if the greatest common divisor of $2p$ and $p^2 + 3q^2$ is 1 (coprime values) and its multiplication is a cube (as we saw in proposition 30.3), then it follows that each of them has to be a cube as well (each k^3 factor of the product, where k is a prime number, has to divide $2p$ or $p^2 + 3q^2$: it is not possible that, for example, k divides $2p$ and k^2 divides $p^2 + 3q^2$, since they are coprime values).

Recall that proposition 30.3 assures us that p and q are both positive integers, of different parity (therefore, $p^2 + 3q^2$ is odd) and coprime. Then, by proposition 30.2, since $p^2 + 3q^2$ is a cube, there exist values t and w such that $p^2 + 3q^2 = (t^2 + 3w^2)^3$, $p = t \cdot (t^2 - 9w^2)$ and $q = 3w \cdot (t^2 - w^2)$. In addition, t and w have different parity to fulfill that $t^2 + 3w^2$ is odd, and $(t, w) = 1$ if we want that $(p, q) = 1$.

Since $p = t \cdot (t^2 - 9w^2)$ then $2p = 2t \cdot (t^2 - 9w^2)$ and, since $2p$ is a cube, then $2t \cdot (t - 3w) \cdot (t + 3w)$ is also a cube. But values $2t$, $t - 3w$, $t + 3w$ are coprime because:

- First, $2t$ is coprime with $t - 3w$ and $t + 3w$, since $t - 3w$ and $t + 3w$ are odd (t and w have different parity) and, in addition, if t has a common factor with any of them then it would also have a common factor with w (which contradicts that $(t, w) = 1$).
- If an odd prime greater than 3 divides $t - 3w$ and $t + 3w$, then it would also divide its sum (that is $2t$) and its subtraction (that is $6w$). But this is impossible since $(t, w) = 1$.
- It only remains to see that 3 does not divide $t - 3w$ and $t + 3w$. But then it would divide its sum $(2t)$, which implies that it would divide t, implying that it would divide p too (since $p = t \cdot (t^2 - 9w^2)$). But that contradicts the fact that $(2p, p^2 + 3q^2) = 1$.

Therefore, if $2p = 2t \cdot (t - 3w) \cdot (t + 3w)$ is a cube and the three factors are coprime numbers, then each of them is a cube and we can write:

$$2t = z_1^3 \qquad t - 3w = x_1^3 \qquad t + 3w = y_1^3$$

But, with these definitions, the equation (30.1) is fulfilled with the values (x_1, y_1, z_1):

$$z_1^3 = 2t = (t - 3w) + (t + 3w) = x_1^3 + y_1^3$$

and, in addition, this solution is **smaller** than (x_0, y_0, z_0) in the sense that the multiplication of the three values yields the inequality:

$$x_1^3 \cdot y_1^3 \cdot z_1^3 = 2p < (2p) \cdot (p^2 + 3q^2) < x_0^3 \cdot y_0^3 \cdot z_0^3$$

where the last inequality is true since we saw in the proof of proposition 30.4 that either $(2p) \cdot (p^2 + 3q^2) = z_0^3$ or $(2p) \cdot (p^2 + 3q^2) = x_0^3$. $\qquad \square$

Since numbers (x_1, y_1, z_1) are different from 0 (by definition) and the cube of their multiplication is smaller than the cube of the multiplication of (x_0, y_0, z_0), then we will find a contradiction repeating this process as many times as necessary (method of infinite descent of natural numbers). Therefore, there cannot be a solution of equation (30.1) when $(2p, p^2 + 3q^2) = 1$.

We have to prove the same when $(2p, p^2 + 3q^2) = 3$.

PROPOSITION 30.6. *Let (x_0, y_0, z_0) be a solution to equation (30.1) with positive coprime values. Let p and q be the numbers calculated as in proposition 30.3. If the greatest common divisor of $2p$ and $p^2 + 3q^2$ is 3, then there is a new solution (x_1, y_1, z_1) to equation (30.1), with positive coprime values too and which also satisfies the property that the product of its cubes $x_1^3 \cdot y_1^3 \cdot z_1^3$ is **smaller** than the product of the cubes of the original values $x_0^3 \cdot y_0^3 \cdot z_0^3$.*

PROOF. The proof is very similar to that of the previous proposition and we are going to give only the main ideas, leaving the details to the reader. The steps to be deducted are the following:

- $(2p, p^2 + 3q^2) = 3$ implies that 3 divides p but not q. Let $s = p/3$; then, $(s, q) = 1$.
- $2p \cdot (p^2 + 3q^2)$, which is a cube, can be written as $18s \cdot (3s^2 + q^2)$. Then, $(18s, 3s^2 + q^2) = 1$ and, therefore, it follows that both $18s$ and $3s^2 + q^2$ are cubes.
- By proposition 30.2, there exist values t and w such that $3s^2 + q^2 = (t^2 + 3w^2)^3$, $q = t \cdot (t^2 - 9w^2)$ and $s = 3w \cdot (t^2 - w^2)$. (Notice the change of role of q from the previous proof).
- That means that $18s$, which is a cube, can be written as $27 \cdot 2w \cdot (t - w) \cdot (t + w)$. Then $2w$, $t - w$ and $t + w$ are coprime values, so each of them is a cube.

- We define $2w = z_1^3$, $t - w = x_1^3$, $t + w = y_1^3$ and we complete the proof similarly to how we did in the previous proposition.

$\square$

We have shown, with a lot of effort but great satisfaction, that no non-trivial solution to Fermat's Theorem for cubes can exist.

FINAL REMARKS

- The error that Euler made was in proposition 30.2 (*"Let $a^2 + 3b^2$ be an odd E-number, not a multiple of 3, with a, b positive, $(a, b) = 1$ and such that it is the cube of an integer, so there exist t and w such that $a^2 + 3b^2 = (t^2 + 3w^2)$, $a = t \cdot (t^2 - 9w^2)$ and $b = 3w \cdot (t^2 - w^2)$"*), which was correctly stated but not completely proved. The ideas for the proof that we have given here (based on the properties of the E-numbers) were also original from Euler, so it is clear that he had the knowledge to correct the proof in case that someone had warned him of his error.

- There are now others proofs for the case $n = 3$ although non-elemental mathematical tools are used in them. Specifically, Gauss provided a proof based on properties of the numbers of the form $a + b \cdot \sqrt{-3}$ (where a and b are rational numbers), which is the one explained in Heinrich Dorrie's book. Other proofs followed later, where more fundamentals of algebra are necessary to understand them, and they are usually more elegant and brief.

- Other cases of Fermat's Theorem were proved later (power-of-2 exponents, $n = 5$, $n = 7$, exponents smaller than 100, etc.), but the general case resisted to be solved and, what is worse, it was anticipated that the complexity of its solution was increasing. Finally, the English mathematician Andrew WILES presented, in 1995, a proof that uses very complex and advanced tools, so only a few dozen people in the world can understand it.

- When in doubt, there will always be romantics who believe that there is a simpler proof to Fermat's Theorem that only he was able to find. However, it is most likely that the proof that he thought he had discovered contained at least one error.